Introduction to Polymer Science

Introduction to Polymer Science

Isaac Fitzgerald

Larsen & Keller
www.larsen-keller.com

Introduction to Polymer Science
Isaac Fitzgerald
ISBN: 978-1-64172-677-1 (Hardback)

Published by Larsen and Keller Education,
5 Penn Plaza,
19th Floor,
New York, NY 10001, USA

Cataloging-in-Publication Data

Introduction to polymer science / Isaac Fitzgerald.
    p. cm.
Includes bibliographical references and index.
ISBN 978-1-64172-677-1
1. Polymers. 2. Polymerization. 3. Macromolecules. I. Fitzgerald, Isaac.
QD381 .I58 2022
547.7--dc23

For more information regarding Larsen and Keller Education and its products, please visit the publisher's website www.larsen-keller.com

Table of Contents

Preface

Polymer science is a subfield of materials science. It generally deals with synthetic polymers such as plastics and elastomers. It has three main sub-disciplines- polymer chemistry, polymer physics and polymer characterization. The chemical synthesis and chemical properties of polymers are studied under polymer chemistry. Polymer physics focuses on the bulk properties of polymer materials and engineering applications. The analysis of chemical structure and morphology is dealt with under polymer characterization. This branch also determines the physical properties with respect to compositional and structural parameters. The various sub-fields of polymer science along with technological progress that have future implications are glanced at in this book. It is appropriate for students seeking detailed information in this area as well as for experts. Coherent flow of topics, student-friendly language and extensive use of examples make this book an invaluable source of knowledge.

To facilitate a deeper understanding of the contents of this book a short introduction of every chapter is written below:

Chapter 1- A large molecule which consists of repeated subunits is known as a polymer. The field of materials science which deals with polymers such as plastics and elastomers is known as polymer science. This chapter has been carefully written to provide an easy understanding of the varied facets of polymer science such as types of polymers and their applications.

Chapter 2- The process through which monomer molecules chemically react to form polymer chains is known as polymerization. The mechanisms of polymerization are broadly divided into two categories, namely, step-growth and chain-growth polymerization. This chapter discusses in detail these mechanisms in related to polymerization.

Chapter 3- The manufacturing activity of transforming raw polymeric materials into final products of desired properties and microstructure are known as polymer processing. Some of the types of polymer processing techniques and processes are molding, extrusion, spinning and vulcanization. The topics elaborated in this chapter will help in gaining a better perspective about these types of polymer processing.

Chapter 4- The polymer which is created when more than one type of polymers are connected in the same polymer chain is known as copolymer. This chapter has been carefully written to provide an easy understanding of the varied facets of copolymerization as well as copolymerization reactions and kinetics.

Chapter 5- The process by which a chemical is converted from a liquid solution into a solid crystalline state is known as crystallization. Polymer crystallization refers to a process which is associated with the partial alignment of the molecular chains in polymers. The topics elaborated in this chapter will help in gaining a better perspective about various aspects of polymer crystallization as well as its kinetics.

Chapter 6- The branch of science which focuses on the study of heat and temperature in polymer solutions is known as thermodynamics. The scientific discipline which studies the deformation of polymeric fluids under external stress is known as polymer rheology. The diverse aspects of thermodynamics of polymer solutions as well the rheological properties of polymers have been thoroughly discussed in this chapter.

I would like to share the credit of this book with my editorial team who worked tirelessly on this book. I owe the completion of this book to the never-ending support of my family, who supported me throughout the project.

Isaac Fitzgerald

Understanding Polymer Science

A large molecule which consists of repeated subunits is known as a polymer. The field of materials science which deals with polymers such as plastics and elastomers is known as polymer science. This chapter has been carefully written to provide an easy understanding of the varied facets of polymer science such as types of polymers and their applications.

Polymer and Polymer Science

Polymer is any of a class of natural or synthetic substances composed of very large molecules, called macromolecules that are multiples of simpler chemical units called monomers. Polymers make up many of the materials in living organisms, including, for example, proteins, cellulose, and nucleic acids. Moreover, they constitute the basis of such minerals as diamond, quartz, and feldspar and such man-made materials as concrete, glass, paper, plastics, and rubbers.

The word *polymer* designates an unspecified number of monomer units. When the number of monomers is very large, the compound is sometimes called a high polymer. Polymers are not restricted to monomers of the same chemical composition or molecular weight and structure. Some natural polymers are composed of one kind of monomer. Most natural and synthetic polymers, however, are made up of two or more different types of monomers; such polymers are known as copolymers.

Natural Rubber: Latex tapped from a rubber tree (Hevea brasiliensis) in Malaysia.

Organic polymers play a crucial role in living things, providing basic structural materials and participating in vital life processes. For example, the solid parts of all plants are made up of polymers.

These include cellulose, lignin, and various resins. Cellulose is a polysaccharide, a polymer that is composed of sugar molecules. Lignin consists of a complicated three-dimensional network of polymers. Wood resins are polymers of a simple hydrocarbon, isoprene. Another familiar isoprene polymer is rubber.

Other important natural polymers include the proteins, which are polymers of amino acids, and the nucleic acids, which are polymers of nucleotides—complex molecules composed of nitrogen-containing bases, sugars, and phosphoric acid. The nucleic acids carry genetic information in the cell. Starches, important sources of food energy derived from plants, are natural polymers composed of glucose.

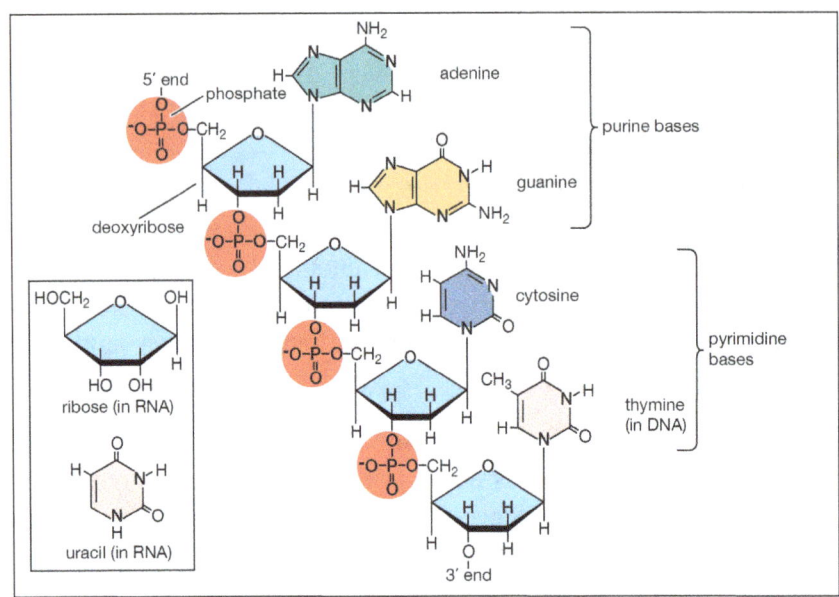

Portion of polynucleotide chain of deoxyribonucleic acid (DNA): The inset shows the corresponding pentose sugar and pyrimidine base in ribonucleic acid (RNA).

Many inorganic polymers also are found in nature, including diamond and graphite. Both are composed of carbon. In diamond, carbon atoms are linked in a three-dimensional network that gives the material its hardness. In graphite, used as a lubricant and in pencil "leads," the carbon atoms link in planes that can slide across one another.

Synthetic polymers are produced in different types of reactions. Many simple hydrocarbons, such as ethylene and propylene, can be transformed into polymers by adding one monomer after another to the growing chain. Polyethylene, composed of repeating ethylene monomers, is an addition polymer. It may have as many as 10,000 monomers joined in long coiled chains. Polyethylene is crystalline, translucent, and thermoplastic—i.e., it softens when heated. It is used for coatings, packaging, molded parts, and the manufacture of bottles and containers. Polypropylene is also crystalline and thermoplastic but is harder than polyethylene. Its molecules may consist of from 50,000 to 200,000 monomers. This compound is used in the textile industry and to make molded objects.

Other addition polymers include polybutadiene, polyisoprene and polychloroprene which are all important in the manufacture of synthetic rubbers. Some polymers, such as polystyrene, are glassy and transparent at room temperature, as well as being thermoplastic. Polystyrene can be coloured any shade and is used in the manufacture of toys and other plastic objects.

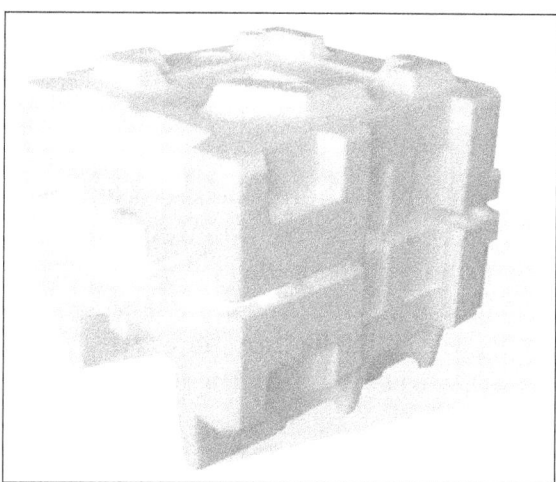

Polystyrene: Polystyrene packaging.

If one hydrogen atom in ethylene is replaced by a chlorine atom, vinyl chloride is produced. This polymerizes to polyvinyl chloride (PVC), a colourless, hard, tough, thermoplastic material that can be manufactured in a number of forms, including foams, films, and fibres. Vinyl acetate, produced by the reaction of ethylene and acetic acid, polymerizes to amorphous, soft resins used as coatings and adhesives. It copolymerizes with vinyl chloride to produce a large family of thermoplastic materials.

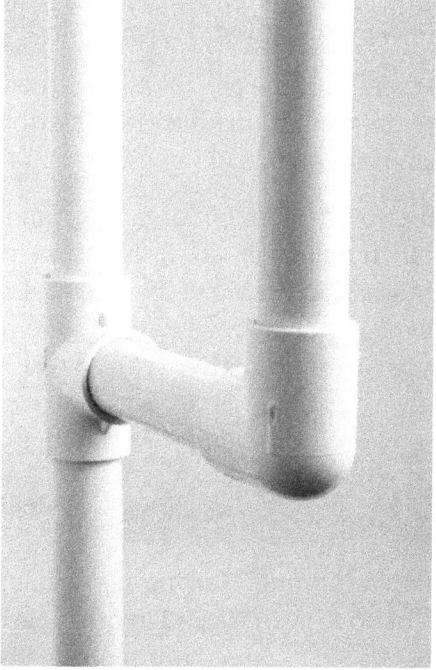

PVC piping: Polyvinyl chloride (PVC) pipes.

Many important polymers have oxygen or nitrogen atoms, along with those of carbon, in the backbone chain. Among such macromolecular materials with oxygen atoms are polyacetals. The simplest polyacetal is polyformaldehyde. It has a high melting point and is crystalline and resistant to abrasion and the action of solvents. Acetal resins are more like metal than are any other plastics and are used in the manufacture of machine parts such as gears and bearings.

A linear polymer characterized by a repetition of ester groups along the backbone chain is called a polyester. Open-chain polyesters are colourless, crystalline, thermoplastic materials. Those with high molecular weights (10,000 to 15,000 molecules) are employed in the manufacture of films, molded objects, and fibres such as Dacron.

The polyamides include the naturally occurring proteins casein, found in milk, and zein, found in corn (maize), from which plastics, fibres, adhesives, and coatings are made. Among the synthetic polyamides are the urea-formaldehyde resins, which are thermosetting. They are used to produce molded objects and as adhesives and coatings for textiles and paper. Also important are the polyamide resins known as nylons. They are strong, resistant to heat and abrasion, noncombustible, and nontoxic, and they can be coloured. Their best-known use is as textile fibres, but they have many other applications.

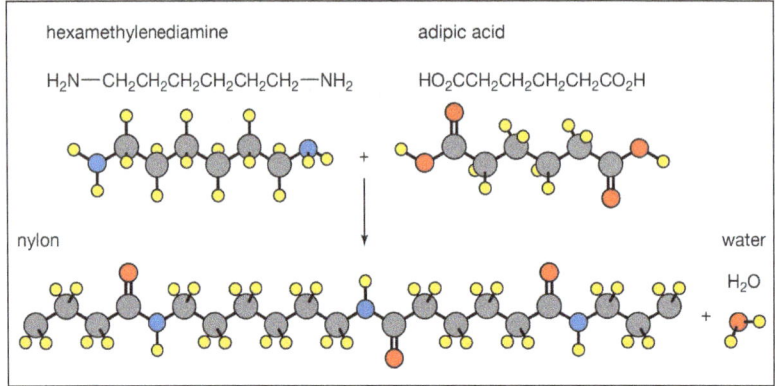

Nylon: The formation of nylon, a polymer.

Another important family of synthetic organic polymers is formed of linear repetitions of the urethane group. Polyurethanes are employed in making elastomeric fibres known as spandex and in the production of coating bases and soft and rigid foams.

A different class of polymers are the mixed organic-inorganic compounds. The most important representatives of this polymer family are the silicones. Their backbone consists of alternating silicon and oxygen atoms with organic groups attached to each of the silicon atoms. Silicones with low molecular weight are oils and greases. Higher-molecular-weight species are versatile elastic materials that remain soft and rubbery at very low temperatures. They are also relatively stable at high temperatures.

Molecular Weight

Let's think about a small molecule, say hexane. Hexane has a molecular weight of 86. Every hexane molecule has a molecular weight of 86. Now if we add another carbon to our chain, and the appropriate amount of hydrogen atoms, we've increased our molecular weight to 100.

$$H-\overset{\overset{\displaystyle H}{|}}{\underset{\underset{\displaystyle H}{|}}{C}}-\overset{\overset{\displaystyle H}{|}}{\underset{\underset{\displaystyle H}{|}}{C}}-\overset{\overset{\displaystyle H}{|}}{\underset{\underset{\displaystyle H}{|}}{C}}-\overset{\overset{\displaystyle H}{|}}{\underset{\underset{\displaystyle H}{|}}{C}}-\overset{\overset{\displaystyle H}{|}}{\underset{\underset{\displaystyle H}{|}}{C}}-\overset{\overset{\displaystyle H}{|}}{\underset{\underset{\displaystyle H}{|}}{C}}-H$$

Hexane has one molecular weight, 86.

$$H-\overset{\displaystyle H}{\underset{\displaystyle H}{C}}-\overset{\displaystyle H}{\underset{\displaystyle H}{C}}-\overset{\displaystyle H}{\underset{\displaystyle H}{C}}-\overset{\displaystyle H}{\underset{\displaystyle H}{C}}-\overset{\displaystyle H}{\underset{\displaystyle H}{C}}-\overset{\displaystyle H}{\underset{\displaystyle H}{C}}-\overset{\displaystyle H}{\underset{\displaystyle H}{C}}-H$$

Lengthening the carbon chain by one carbon
turns hexane into a completely different
compound, heptane, molecular wieght = 100.

That's fine, but the molecule is no longer hexane. It's heptane! If we have a mixture of some molecules of hexane and some of heptane, the mixture won't act like pure heptane, nor will it act like pure hexane. The properties of the mixture, say, its boiling point, vapor pressure, etc., will be neither those of pure hexane nor pure heptane.

But polymers are different. Imagine polyethylene. If we have a sample of polyethylene, and some of the chains have fifty thousand carbon atoms in them, and others have fifty thousand and two carbon atoms in them, this little difference isn't going to amount to anything. If you really want to know the truth, one almost never finds a sample of a synthetic polymer in which all the chains have the same molecular weight. Instead, we usually have a bell curve, a *distribution* of molecular weights. Some of the polymer chains will be much larger than all the others, at the high end of the curve. Some will be much smaller, and at the low end of the curve. The largest number will usually be clumped around a central point, the highest point on the curve.

So we have to talk about *average* molecular weights when we talk about polymers. And we're not going to stop there. The average can be calculated in different ways, and each way has its own value.

Number Average Molecular Weight, Mn

The number average molecular weight is not too difficult to understand. It is just the total weight of all the polymer molecules in a sample, divided by the total number of polymer molecules in a sample.

Weight Average Molecular Weight, Mw

The weight average is a little more complicated. It's based on the fact that a bigger molecule contains more of the total mass of the polymer sample than the smaller molecules do.

Demographics

A good way to understand the difference between the number average molecular weight and the weight average molecular weight is to compare some American cities.

Let's take four cities, say, Memphis, Tennessee; Montrose, Colorado; Effingham, Illinois; and Freeman, South Dakota. Now we'll take a look at their populations:

City	Population
Memphism, Tennessee	700,000
Montrose, Colorado	10,000
Effingham, Illinois	12,000
Freeman, South Dakota	1,500

Now let's calculate the simple average population of the four cities:

$$700,00$$
$$10,000$$
$$12,000$$
$$+\quad 1,500$$
$$\overline{\hspace{1cm}}$$
$$723,500$$

$$\frac{723,500}{4}=180875$$

Wow!

Now we see that of these four cities, that average population is 180,875.

But we could look at it a different way. Until now we've been worried about "the average city". What is the population of "the average city"? But let's forget about cities for a moment, and think about people. What size city does the average person among the populations of these four towns live in? If you look at the numbers you can see that the average person doesn't live in a town of a population of 180,000. Take a look there. Most of the people in the combined populations of the four towns live in Memphis, a town with a lot more than 180,000 people. So how do we calculate the size of town that the average person lives in, if the simple average doesn't work? What we need is a weighted average. This is an average that would account for the fact that a large city like Memphis holds a larger percentage of the total population of the four cities than Montrose, Colorado. Doing this involves a little bit of math that looks scary but really isn't. All we do is take the total number of people in each city, then multiply that number by that city's fraction of the total population. Take all the answers we get for each city and add them up, and we get an answer that we'll call the weight average population of the four cities.

Take Memphis. It has a population of 700,000. The total population of our four cities is 723,500.

$$\frac{700,000}{723,500}=0.9675.$$

So the fraction of people who live in Memphis is 0.9675, or we might say, 96.75% of the people live in Memphis. Now let's take our fraction, 0.9675, and multiply that by the population of Memphis:

$$700,000 \times 0.9675 = 677,263.3$$

And we get an answer of 677,273.3. Now let's do the same thing for all the cities, and add up the answers:

$$700,000 \times \frac{700,000}{723,500} = 700,000 \times 0.9675 = 677,263.3$$

$$10,000 \times \frac{10,000}{723,500} = 10,000 \times 0.0138 = 138.2$$

$$12,000 \times \frac{12,000}{723,500} = 12,000 \times 0.0166 = 199.0$$

$$1,500 \times \frac{1,500}{723,500} = 1,500 \times 0.00207 = 3.2$$

$$\overline{677,603.7}$$

weight average population is around 677,600.

So our weight average population of the four cities is about 677,600. We can say from this figure that the average person lives in a city of about 677,600. That is more believable than saying that the average citizen lives in a city of 180,000.

We do the same thing with polymers. We calculate, with the same formula as we used for the weight average population of our four cities, the weight average molecular weight.

Viscosity Average Molecular Weight, Mv

Molecular weight can also be calculated from the viscosity of a polymer solution. The principle is a simple one: Bigger polymers molecules make a solution more viscous than small ones do. Of course, the molecular weight obtained by measuring the viscosity is a different from either the number average or the weight average molecular weight. But it's closer to the weight average than the number average.

Distribution

With all these different molecular weights out there, things can get a little confusing. No single one of them tells the whole story. So it's usually best to try to know the molecular weight distribution. The distribution is a plot, like the one in the picture. It plots molecular weight on the x-axis, and plots the amount of polymer at a given molecular weight on the y-axis. Just for fun, we've shown you just where on the distribution curve the number, viscosity, and weight averages generally show up.

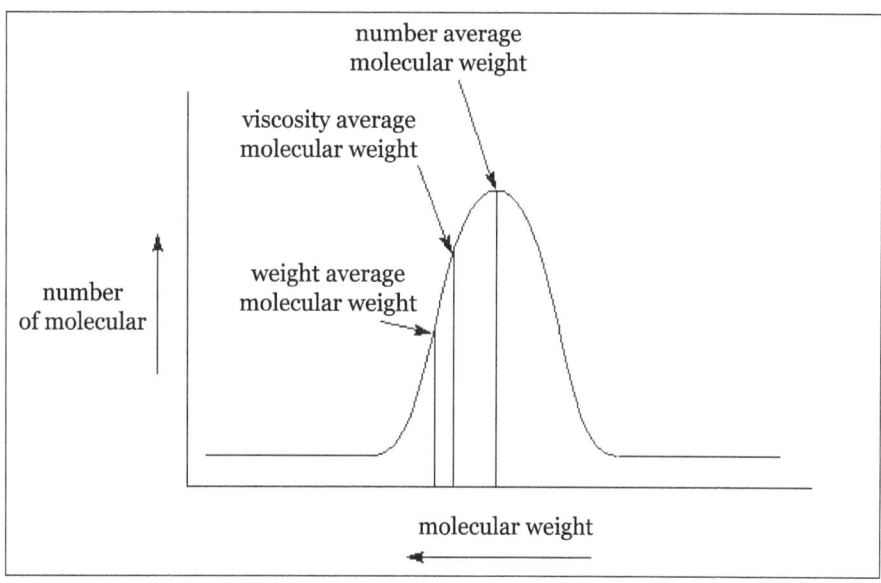

Renegade Distributions

If we lived in a perfect world, where molecular distributions were always so nice and bell shaped, just knowing the averages might be enough. But they aren't always like that. Sometimes they are like this:

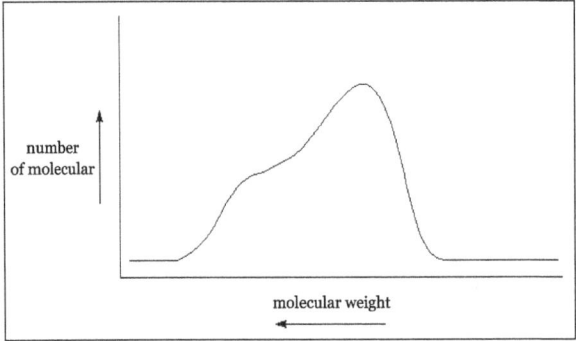

This kind of distribution can result from something called a Tromsdorff effect, which we find in free radical vinyl polymerization. Sometimes the distribution is even nastier, like this:

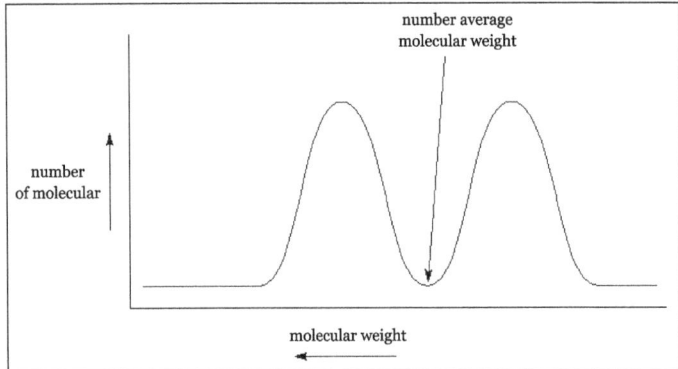

Here our number average molecular weight is a complete lie. There isn't a single molecule of that weight in the whole sample! Cases like these illustrate the need to know the complete distribution. The distribution can be given by a technique called size exclusion chromatography, and also by a new method called MALDI mass spectrometry.

Weight Distribution

The polymerization process, whether proceeding by chain growth or by step-growth, is ruled by random events. The result is a mixture of polymers that vary in chain length. A polymeric material, therefore, cannot be characterized by a single molecular weight like an ordinary substance. Instead, a statistical average calculated from the molecular weight distribution has to be used.

The average can be expressed in two ways. One way is to calculate the number average, which is the sum of all molecular weights divided by their total number:

$$\overline{M_n} = \frac{\sum_{i=1}^{N} N_i M_i}{\sum_{i=1}^{N} N_i} = \frac{\sum_{i=1}^{N} W_i}{\sum_{i=1}^{N} W_i / M_i} = \frac{1}{\sum_{i=1}^{N} w_i / M_i}$$

Where N is the total number of molecules, N_i the number of molecules having a molecular weight M_i, and w_i is the weight fraction of all molecules having a molecular weight M_i.

Another way to express the average molecular weight is to calculate the weight average, which is the sum of all molecular weights multiplied by their weight fractions:

$$\overline{M_w} = \frac{\sum_{i=1}^{N} N_i M_i^2}{\sum_{i=1}^{N} N_i M_i} = \frac{\sum_{i=1}^{N} W_i M_i}{\sum_{i=1}^{N} W_i} = \sum_{i=1}^{N} w_i M_i.$$

The two expressions for the average molecular weight are special cases of the general expression for weight averages:

$$\overline{M_z} = \frac{\sum_{i=1}^{N} N_i M_i^{\alpha}}{\sum_{i=1}^{N} N_i M_i^{\alpha-1}} = \frac{\sum_{i=1}^{N} W_i M_i^{\alpha-1}}{\sum_{i=1}^{N} W_i M_i^{\alpha-2}}.$$

The parameter α is the so-called weighting factor, which defines the particular average. The higher averages, which are often called z-averages, are more sensitive to high molecular weight portions and are more difficult to measure accurately. They are related to methods that measure the motion of polymer molecules, such as diffusion or sedimentation methods.

It can be shown that the weight average molecular weight is a good measure for the expected statistical size of the polymer, whereas the number average molecular weight is a measure for the chain length. The two averages can lead to very different molecular weight averages. The weight average is particularly sensitive to the presence of higher molecular weight molecules whereas the number average is very sensitive to the presence of lower molecular weight molecules. For example, if equal parts by weight of molecules with a molecular weight of 10,000 and 100,000 g/mol are mixed then the weight average molecular weight is 55,000 g/mol whereas the number average is only 18182 g/mol. If, on the other hand, equal numbers of both molecules are mixed then the weight average is 91818 g/mol and the number average 55,000 g/mol. For all polydisperse synthetic polymers with bell-shaped distribution of molecular weight we find:

$$M_n < M_w < M_z < M_{z+1}$$

The ratio M_w/M_n is called the *polydispersity* or *heterogeneity index*. It is a measure for the broadness of a molecular weight distribution of a polymer, that is, the larger the polydispersity index, the broader the molecular weight distribution.

The average molecular weight is related to the viscosity of the polymer under specific conditions. In the case of solution viscosity, the weight dependence of the viscosity can be described by the well-known empirical Mark-Houwink relation:

$$[\eta] = K_{\eta} M_{\eta}^{\alpha}$$

where $[\eta]$ is the intrinsic viscosity, and α, K_{η} are the Mark-Houwink parameters. These two quantities have been measured for many polymers.

Measurements of the viscosity yield the viscosity average molar weight:

$$\overline{M_\eta} = \left[\frac{\sum_{i=1}^{N} N_i M_i^{\alpha+1}}{\sum_{i=1}^{N} N_i M_i} \right]^{1/\alpha} = \left[\frac{\sum_{i=1}^{N} W_i M_i^{\alpha}}{\sum_{i=1}^{N} W_i} \right] = \left[\sum_{i=1}^{N} w_i M_i^{\alpha} \right]^{1/\alpha}$$

The viscosity average is usually larger than the mass average but smaller than the number average, $M_n < M_\eta < M_w$. Two very common techniques for measuring the molecular mass of polymers are high-pressure liquid chromatography (HPLC), also known as size exclusion chromatography (SEC), and gel permeation chromatography (GPC). These techniques are based on forcing a polymer solution through a matrix of cross-linked polymer particles at high pressure of up to several hundred bars.

Effect of Molecular Weight, Dispersity and Branching on Polymer Properties

The molecular weight, dispersity and branching has a significant effect on the mechanical and physical bulk properties of polymers. In general, a higher molecular weight improves the mechanical properties, that is, break, yield, and impact strength increase. However, a higher molecular weight also increases the melt and glass transitions temperature as well as the solution and melt viscosity which makes processing and forming of the polymeric material more difficult.

The *dispersity* has the opposite effect; a wider molecular weight distribution lowers the tensile and impact strength but increases the yield strength, or in other words, a lower dispersity (narrower distribution) leads to better mechanical properties. The low-molecular weight portion of the distribution has a similar effect as a plasticizer, that is, it reduces the brittleness and lowers the melt viscosity which improves the processability, whereas the high-molecular weight portion causes processing diffculties because of its huge contribution to the melt viscosity.

Branching is another important performance parameter. In general, branching lowers the mechanical properties. For example, it decreases the break and yield strength. The effect on toughness is less clear; if the length of the branches exceeds the entanglement weight it improves the toughness, otherwise it lowers the impact strength. Branching also lowers the brittleness, the melt temperature the melt and solution viscosity and increases the solubility. In conclusion, the processability improves with increasing degree of branching.

Types of Polymers

A polymer is a large molecule that is made up of repeating subunits connected to each other by chemical bonds. Do you need some examples of polymers? Here is a list of materials that are natural and synthetic polymers, plus some examples of materials that are not polymers at all.

Natural Polymers

Polymers are both found in nature and manufactured in laboratories. Natural polymers were used for their chemical properties long before they were understood in the chemistry laboratory: Wool, leather, and flax were processed into fibers to make clothing; animal bone was boiled down to make glues. Natural polymers include:

- Proteins, such as hair, nails, tortoiseshell.

- Cellulose in paper and trees.

- Starches in plants such as potatoes and maize.

- DNA.

- Pitch (also known as bitumen or tar).

- Wool (a protein made by animals).

- Silk (a protein made by insects).

- Natural rubber and lacquer (proteins from trees).

Synthetic Polymers

Polymers were first manufactured by people seeking substitutes for natural ones, in particular, rubber and silk. Among the earliest was semi-synthetic polymers, which are natural polymers modified in some way. By 1820, natural rubber was modified by making it more fluid; and cellulose nitrate prepared in 1846 was used first as an explosive and then as a hard moldable material used in collars, Thomas Edison's film for movies and Hilaire de Chardonnet's artificial silk (called nitrocellulose).

Fully synthetic polymers include:

- Bakelite, the first synthetic plastic.

- Neoprene (a manufactured form of rubber).

- Nylon, polyester, rayon (manufactured forms of silk).

- Polyethylene (plastic bags and storage containers).

- Polystyrene (packing peanuts and Styrofoam cups).

- Teflon.

- Epoxy resins.

- Silicone.

- Silly putty.

- Slime.

Non-polymers

So while paper plates, styrofoam cups, plastic bottles, and a block of wood are all examples of polymers, there are some materials which are *not* polymers. Examples of materials which are not polymers include:

- Elements.

- Metals.

- Ionic compounds, such as salt.

Usually, these materials form chemical bonds, but not the long chains that characterize polymers. There are exceptions. For example, graphene is a polymer made up of long carbon chains.

Bonding of Polymer Materials

Metals, polymers and ceramics have contrasting physical and chemical properties. Cotton t-shirts and plastic spoons do not have much in common with bicycle frames or coffee mugs. Polymers have low densities; do not reflect or absorb light (they are white or colorless); do not conduct electricity; and are flammable.

A primary reason that polymer properties are different is because the chemical compositions of metals, polymers and ceramics are totally different. Polymers are composed of non-metallic elements, found at the upper right corner of the periodic table. Carbon is the most common element in polymers. The chemical bonds in polymers are also different than those found in metals and ceramics.

Covalent Bonds

Non-metallic elements have a high number of valence electrons (four or more) and prefer to gain electrons, not lose them, in chemical reactions. They often form anions. In a compound of only nonmetals there are no elements willing to become cations so ionic bonds are not possible. Instead, two nonmetallic atoms can share valence electrons with each other. This type of electron sharing, called covalent bonding, keeps the shared electrons close to both atomic nuclei. One pair of shared electrons makes one covalent bond. A molecule is a group of atoms held together by covalent bonds.

This type of bonding contrasts with metallic bonding, in which valence electrons are not associated with a particular nucleus, and move easily throughout a sample.

Each hydrogen atom has one proton in its nucleus, and one electron.
The electron moves randomly in a spherical space around the nucleus.

To determine how many covalent bonds can be formed between atoms, first the number of valence electrons must be counted. This can be determined by using a periodic table. The group number matches the number of valence electrons.

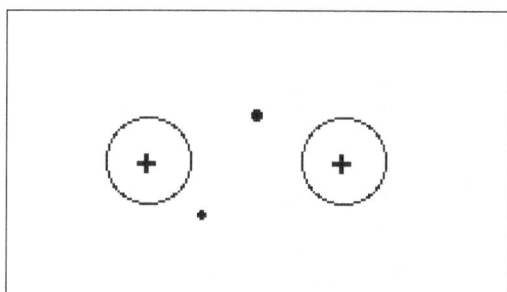

When a covalent bond forms, hydrogen nuclei move close toghether. Although the electrons still move aro9und the molecule, usually they are near both nuclei.

Example: Carbon is element 6. It is found in Group IV so has four valence electrons. Oxygen is element 8. It is found in Group VI so has six valence electrons. The molecule carbon dioxide has the chemical formula CO_2. 4 electrons from C + 2(6) electrons from O = 16 valence electrons.

Lewis Structures

Rather than writing a sentence for the number of valence electrons on an atom, it can be more useful to draw a picture containing this information. A Lewis structure for an atom starts with a chemical symbol, with a dot added for each valence electron.

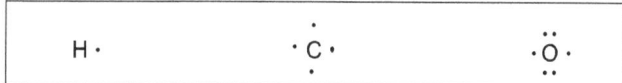

Since the highest possible number of valence electrons is eight (for noble gases) the dots representing valence electrons are traditionally arranged on four sides of the symbol, with at most two electrons on each side.

Experiments have shown that most nonmetallic nuclei are satisfied when they are near eight valence electrons. That is known as the octet rule. Carbon has four valence electrons, and needs to find four more to share. Oxygen has six valence electrons, so it only needs two more. Hydrogen is an exception to the octet rule; its nearest noble gas, helium, has only two electrons. Hydrogen nuclei form molecules with two nearby electrons, a duet rule.

To show a covalent bond, two chemical symbols are put near each other with two dots, representing a pair of electrons, between them. For example, a water molecule has one oxygen atom covalently bound to two hydrogen atoms.

Nuclei do not have to share all of their valence electrons. Note that two pairs of the oxygen's valence electrons are not shared with any other atoms. Those electrons are called lone pairs or nonbonded pairs and can influence the chemical properties of a molecule.

Because dots can be difficult to see, it is common to draw a line segment for each bond (two elec- trons). The Lewis structure of a water molecule then looks like:

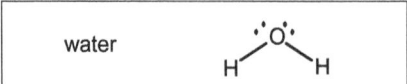

Hydrogen peroxide is a different compound of hydrogen and oxygen, with chemical formula H_2O_2. Since hydrogen atoms only need a duet of electrons they are found at the outside of the molecule; the two oxygen atoms need to be in the center where they can form more bonds. In this textbook that structural information will be given by underlining the central atom(s) in the chemical for- mula: $H_2\underline{O}_2$.

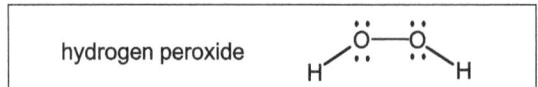

Sometimes nuclei will need to share more than one pair of electrons to achieve an octet with the available valence electrons. One shared pair of electrons is a single bond. Two shared pairs (four electrons) make a double bond, and three shared pairs (six electrons) makes a tri- ple bond. The more electrons that are shared, the stronger the bond will be. The neighboring elements carbon, nitrogen and oxygen commonly use double and triple bonds. For example, both nitrogen and oxygen are usually found as diatomic gases, N_2 and O_2. The nitrogen mol- ecule has $2 \times 5 = 10$ valence electrons, and oxygen molecule has $2 \times 6 = 12$. Nitrogen needs a triple bond to achieve octets for each atom, but a double bond is sufficient for the oxygen molecule.

N≡N   :Ö=Ö:

The type of covalent bond affects the shape of a molecule. The nuclei move closer together if they share more electrons. This means that a triple bond is shorter than a double bond, which is short- er than a single bond. The bond angles are also very specific in a covalently bond molecule. The shared electrons want to be near the two positively charged nuclei, but try to stay away from neg- atively charged lone pairs. This structure is quite different than that created by metallic bonds, which do not have a particular orientation. Covalent molecules can flex a bit under stress but prefer to "bounce back" to their original positions.

Although a Lewis structure is a good way to show covalent bonds between atoms, it is not as ef- fective at showing a molecules' three dimensional shape. Chemists use model kits and chemical graphics programs to visualize the positions of atoms in molecules.

Functional Groups in Polymers

Carbon is the most important element in polymers. Because it starts with only four valence elec- trons, and wants to share four more, carbon forms a wide variety of covalent bonds. Most impor- tantly, carbon forms strong bonds with itself. Long, strong chains or nets made of thousands of carbon atoms form the backbone of a polymer.

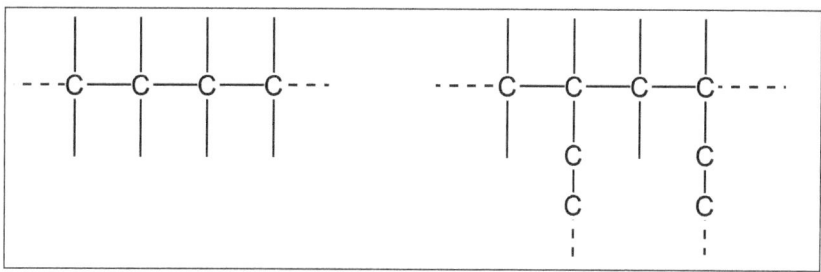

Carbon Backbones

Polyethylene is the simplest polymer. In addition to the carbon backbone, only hydrogen atoms are used to achieve four covalent bonds per carbon atom.

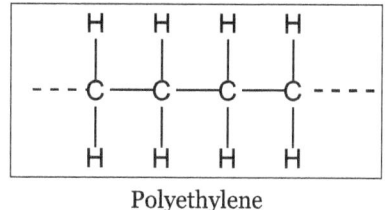

Polyethylene

Although silicon is in the same group as carbon, it does not form strong bonds with itself. Silicones, long chains of alternating silicon and oxygen atoms, can be synthesized.

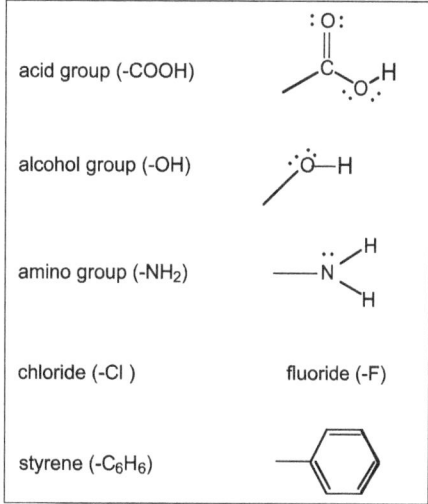

Silicone

Many different nonmetal atoms could be covalently attached to a polymer backbone. Groups of atoms that contribute something besides C-C and C-H bonds are called functional groups. They affect the chemical and physical properties of a polymer.

Examples of Functional Groups:

acid group (-COOH)	
alcohol group (-OH)	
amino group (-NH$_2$)	
chloride (-Cl)	fluoride (-F)
styrene (-C$_6$H$_6$)	

Skeleton Structures

Simplified or "skeleton" structures can be used to emphasize the functional groups. Carbon-carbon bonds of the framework are represented by line segments. Each vertex is the location of a carbon atom. Most hydrogen atoms and all lone pairs are omitted. This type of diagram deemphasizes the hydrocarbon skeleton; since it is so strongly bonded as to be unreactive, it does not affect the chemical properties of the polymer.

Polyethylene is the simplest polymer. Since it has no functional groups, the skeleton structure of a polyethylene fragment looks like it does not have any atoms! (Remember that a real polyethylene molecule is more often 100 or 1000 atoms long).

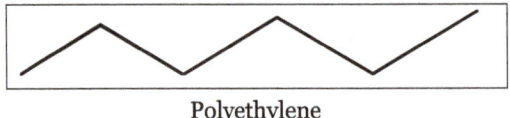

Polyethylene

It is possible to figure out the missing information. There should be a carbon atom at the end of each line segment; six are needed, connected by five single bonds.

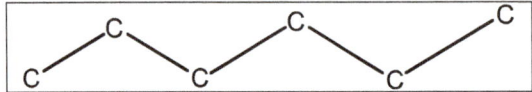

Since each carbon atom must have four bonds in a molecule, there must be missing bonds to hydrogen atoms. For the carbon atoms on the ends of the molecule, adding three C-H bonds to each will achieve octets. Two C-H bonds should be added to each of the inner carbons.

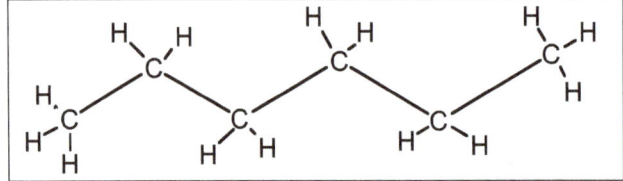

Complete Lewis Structure for Polyethylene Fragment.

Three Dimensional Model of Polyethylene.

When functional groups are added to a simplified backbone it is easy to notice the change in structure. Polyfluoroethylene, often sold as Teflon, is similar in structure to polyethylene except that all the hydrogen is replaced with fluorine. It is a very slippery polymer.

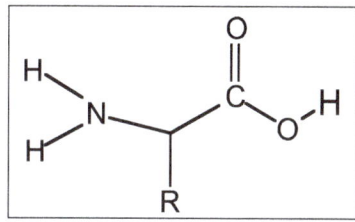

polyfluoroethylene

Amino acids, the monomers that build proteins, contain amino groups and acid groups, separated by one carbon. In the model nitrogen atoms are blue and oxygen atoms are red. The carbon between the amino group and the acid group always has hydrogen on it (pointing up in the model) but the fourth group is variable. The symbol R is used when the exact identity of the group is not important. More than twenty different groups, as simple as a single hydrogen atom, are found on amino acids in nature.

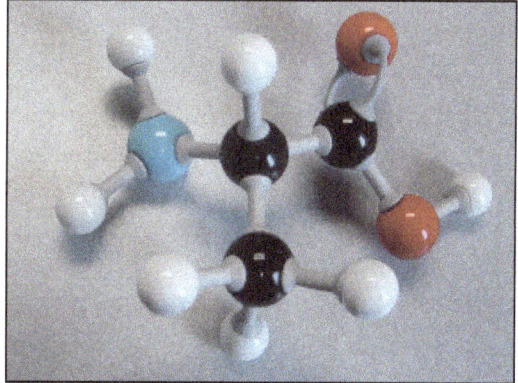

Amino Acid

The photo shows a model of alanine, which has a -CH$_3$ group:

Three Dimensional Model of Amino Acid Alanine.

Polysaccharides are sugar polymers. Cellulose (found in wood, cabbages, cotton, and linen) is composed of long chains of sugar rings. They are covered with alcohol groups.

Cellulose

Starch contains the same functional groups, but the sugar rings are connected at different angles. The structural change makes it possible for the human digestive system to digest the polymer into sugar. Starch polymers are often branched.

Starch

Intermolecular Forces

A molecule is a group of atoms connected by covalent bonds. Chemical reactions are required to form or break covalent bonds. Weaker attractions often form between molecules, encouraging them to stick together in groups. The weaker attractions are called secondary bonds or intermolecular forces. These can be overcome by adding heat or dissolving in a liquid. The functional groups on a polymer determine the types and strength of its secondary bonds.

Polar Interactions

The valence electrons moving around a molecule may not be symmetrically distributed. The nonmetallic elements closest to the right top corner of the periodic table - nitrogen, oxygen, fluorine and chlorine - tend to shift shared electrons away from carbon and hydrogen. When there is a functional group with one of those elements, it has a slight negative charge and the rest of the molecule (carbon and hydrogen) is slightly positive. The molecule is polarized (or polar, for short). Its positive sections are attracted to negative sections of neighboring polymers.

Poly (Ethylene Terephthalate) "PET".

Poly (ethylene terephthalate) or PET, a polymer used to make bottles for carbonated beverages, has oxygen-containing functional groups that make it polar. Protein and cellulose chains are also polars.

Polyfluoroethylene is nonpolar (not polar) because it is completely covered with fluorine atoms; there is no exposed positive section to interact with a neighboring molecule's negative section.

Positive and negative charges can be localized on a covalent molecule since they have no path for conduction of electrons. The carbon atoms in the backbone always follow the octet rule with four covalent bonds, so can't pass extra electrons along the chain. If polymer fibers are rubbed together they can build up a static electricity charge.

Hydrogen Bonds

Molecules with either -N-H or -O-H groups will form strong secondary bonds. This phenomenon is responsible for the relatively high boiling point of water, and for the fact that its solid form (ice) is less dense than its liquid form. Polymers with hydrogen-bonding groups will soak up water.

Fabric softeners are added to laundry to change the properties of cotton and linen fabrics. The fabric softener molecules have one end that binds to OH groups on cellulose. The other end of the fabric softener is a long, nonpolar chain. This exposed end feels smooth and slippery. Softened fabrics are less able to build up a static charge. However softened fabrics will not absorb as much water. This is an issue for the performance of cotton towels.

Nonpolar Interactions

As valence electrons move around the nuclei in a nonpolar polymer, like polyethylene or polyfluoroethylene, they can become temporarily imbalanced. For a brief moment of time one part of a molecule would be negative, another part positive; it is temporarily polar. These occasional imbalances are enough to allow nonpolar molecules to attract each other, but the interaction is much weaker than that observed for polar or hydrogen bonding polymers.

Applications of Polymers

Polymer testing and consultancy for plastics, additives with applications including aerospace, automotive, electronics, packaging and medical devices. Polymers are a highly diverse class of materials which are available in all fields of engineering from avionics through biomedical applications, drug delivery system, biosensor devices, tissue engineering, cosmetics etc. and the improvement and usage of these depends on polymer applications and data obtained through rigorous testing. The applications of polymeric materials and their composites are still increasing rapidly due to their below average cost and ease of manufacture. When considering a polymer application, understanding how a material behaves over time allows us to assess its potential application and use. We can provide failure analysis of polymers and plastics and identify design faults or moulding issues. Our expertise can be applied to simple packaging films all the way through to advanced aerospace materials, and can be used as part of complex litigation cases. Polymeric materials tested include raw materials, polymer compounds, foams, structural adhesives and composites, fillers, films, membranes, emulsions, coatings, rubbers, sealing materials, adhesive resins, solvents, inks and pigments.

Polymers are generally used in the areas of:

- In aircraft, aerospace, and sports equipment.

- Printed circuit board substrates.

- 3D printing plastics.

- Polymers in holography.

- Biopolymers in molecular recognition.

- Polymers in bulletproof vests and fire-resistant jackets.

- Organic polymer flocculants in water purification.

- Green Chemicals: Polymers and Biopolymers.

- Polymeric Biomolecules.

- Monomeric Units.

- Renewable Biomass Sources.

References

- Polymer, science: britannica.com, Retrieved 4 February, 2019

- Weight: pslc.ws, Retrieved 12 April, 2019

- Molecular-Weight, polymer-physics: polymerdatabase.com, Retrieved 18 June, 2019

- What-are-examples-of-polymers-604299: thoughtco.com, Retrieved 8 January, 2019

- Materials: uwosh.edu, Retrieved 10 May, 2019

- Applications-of-polymers: conferenceseries.com, Retrieved 7 March, 2019

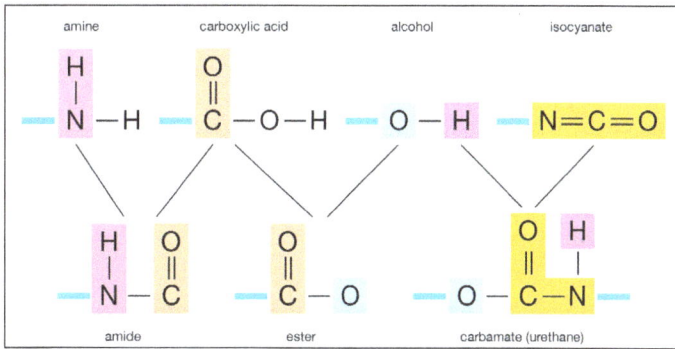

Polymerization and Polymer Synthesis

The process through which monomer molecules chemically react to form polymer chains is known as polymerization. The mechanisms of polymerization are broadly divided into two categories, namely, step-growth and chain-growth polymerization. This chapter discusses in detail these mechanisms related to polymerization.

Polymerization and its Process

Polymerization is any process in which relatively small molecules, called monomers, combine chemically to produce a very large chainlike or network molecule, called a polymer. The monomer molecules may be all alike, or they may represent two, three, or more different compounds. Usually at least 100 monomer molecules must be combined to make a product that has certain unique physical properties—such as elasticity, high tensile strength, or the ability to form fibres—that differentiate polymers from substances composed of smaller and simpler molecules; often, many thousands of monomer units are incorporated in a single molecule of a polymer. The formation of stable covalent chemical bonds between the monomers sets polymerization apart from other processes, such as crystallization, in which large numbers of molecules aggregate under the influence of weak intermolecular forces.

Two classes of polymerization usually are distinguished. In condensation polymerization, each step of the process is accompanied by the formation of a molecule of some simple compound, often water. In addition polymerization, monomers react to form a polymer without the formation of by-products. Addition polymerizations usually are carried out in the presence of catalysts, which in certain cases exert control over structural details that have important effects on the properties of the polymer.

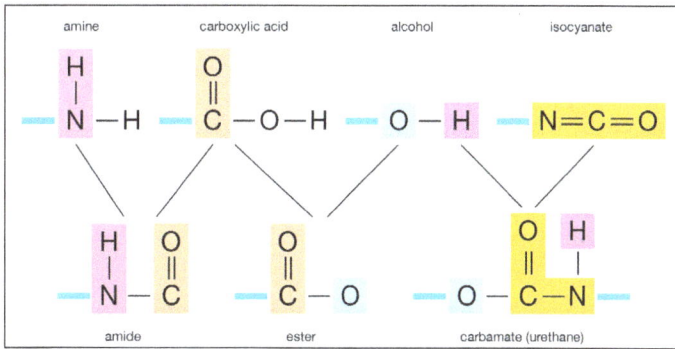

Functional Group: Monomers and Polymers

Linear polymers, which are composed of chainlike molecules, may be viscous liquids or solids with varying degrees of crystallinity; a number of them can be dissolved in certain liquids, and they soften or melt upon heating. Cross-linked polymers, in which the molecular structure is a network, are thermosetting resins (i.e., they form under the influence of heat but, once formed, do not melt or soften upon reheating) that do not dissolve in solvents. Both linear and cross-linked polymers can be made by either addition or condensation polymerization.

Degree of Polymerization

A degree of polymerization is a key characteristic of polymers that determine physical properties of polymer materials. Polymers are large molecules that consist of repeating structural (monomer) units. For example, polyethylene is composed of repeating units $(CH_2-CH_2)_n$ where "n" is an integer number that indicates the degree of polymerization. Mathematically, this parameter is a ratio of the molecular weights of the polymer and the respective monomer unit.

1. Write down the Chemical Formula: Write down the chemical formula of the polymer For example, if the polymer is tetrafluoroethylene then its formula is $-(CF_2-CF_2)_n-$. The monomer unit is placed in parentheses.

2. Get the Atomic Masses: Obtain atomic masses of the elements that compose the monomer unit molecule, using the periodic table of elements. For tetrafluoroethylene, the atomic masses of carbon (C) and fluorine (F) are 12 and 19, respectively.

3. Calculate Molecular Weight: Calculate the molecular weight of the monomer unit by multiplying the atomic mass of each element by the number of atoms in the monomer of each, then add the products. For tetrafluoroethylene, the molecular weight of the monomer unit is $12 \times 2 + 19 \times 4 = 100$.

4. Divide to Get Degree of Polymerization: Divide the molecular weight of the polymer by the molecular weight of the monomer unit to calculate the degree of polymerization. If the molecular mass of tetrafluoroethylene is 120,000, its degree of polymerization is $120,000 / 100 = 1,200$.

Step Growth Polymerization

In a step-growth polymerization, the molecular weight of the polymer chain builds up slowly and there is only one reaction mechanism for the formation of polymer. The distinct initiation, propagation, and termination steps of chain-growth polymerization are meaningless in step-growth polymerization. A difunctional monomer or equal molar amounts of two different difunctional monomers are necessary at least to form a linear high molecular weight polymer. The polymerization reaction proceeds by individual reactions of the functional groups on the monomers. Thus, two monomers react to form a dimer. The dimer may now react with another dimer to produce a tetramer, or the dimer may react with more monomer to form a trimer. This process continues, each reaction of the functional groups proceeding essentially at the same reaction rate until over a relatively long period of time, a high molecular weight polymer is obtained.

Whereas the reaction of two monofunctional compounds, hexanoic acid and n-hexylamine, gives one amide of molecular weight of 199, the reaction of two difunctional monomers, adipic acid and hexamethylene diamine can produce a high molecular weight, (~25,000) poly(hexamethylene-adipamide). As an alternative, e-aminocaproic acid, a single difunctional monomer, can also give a high molecular weight polyamide.

$$n \ \text{HO-}\overset{\overset{\displaystyle O}{\|}}{\text{C}}\text{-(CH}_2)_4\ \overset{\overset{\displaystyle O}{\|}}{\text{C}}\text{-OH} + n \ \text{H}_2\text{N(CH}_2)_6\ \text{NH}_2 \rightarrow$$

$$\left[\overset{\displaystyle H}{\underset{\displaystyle}{\text{N}}}\text{-}\overset{\overset{\displaystyle O}{\|}}{\text{C}}\text{(CH}_2)_4\ \overset{\overset{\displaystyle O}{\|}}{\text{C}}\text{- N(CH}_2)_6 \right]_n + 2n \ \text{H}_2\text{O}$$

$$n\text{H}_2 \ \text{N(CH}_2)_5\ \overset{\overset{\displaystyle O}{\|}}{\text{C}}\text{-OH} \rightarrow \left[\text{(CH}_2)_5\ \overset{\overset{\displaystyle O}{\|}}{\text{C}}\text{- N}\overset{\displaystyle H}{} \right]_n + n \ \text{H}_2\text{O}$$

Some of the similarities and differences in the step-growth and chain-growth polymerization are listed in the table.

Requirements for High Molecular Weight

There are three critical requirements for the step-growth polymerization to yield a high molecular weight linear polymer. First, a perfect stoichiometric balance of the two difunctional monomers must be introduced, or alternately a self-balancing reaction is necessary. Of course, when a single difunctional monomer can generate polymer, such as is the case of f-aminocaproic acid, an internal balance (within the monomer) is provided.

Second, a high degree monomer purity is necessary. It is evident that in the case of e-aminocaproic acid, if the decarboxylation product, n -pentylamine, is present, then internal balance is no longer achieved. Likewise, the presence of a trace of monocarboxylic acid ir) adipic will lead to a stoichiometric imbalance. Furthermore, these mono-functional monomers act as caps to the polymer chain. Once the monocarboxylic acid has undergone amide formation, no further reaction is possible at that end of the chain.

Third, the reaction responsible for the polymerization must be a very high yield reaction with the absence of side reactions. Of the large number of reactions known to organic chemists today, only four are utilized in the synthesis of step-growth polymers in large amounts. These four reactions meet the requirements of high yield reactions and cost feasibility.

1. Ester interchange.
2. Amidation by dehydration of an ammonium salt.
3. Reaction of an isocyanate with an alcohol (and amine).
4. A Schotten Baumann reaction of an acid chloride with an amine (or alcohol).

These requirements apply to the formation of high molecular weight linear polymers from difunctional monomers. If the monomer functionality is greater than two, then these requirements do not apply, and the resulting polymer is either highly branched or a cross-linked, three-dimensional network.

Let us examine these three criteria in the reverse order by first considering the relationship between the yield or the extent of a reaction and the molecular weight of the polymer.

Step-growth vs. Chain-growth Polymerization

	Step-Growth	Chain-Growth
• Reactions	One reaction is responsible for polymer formation.	Initiation, propagation, and termination reactions have different rates and mechanisms.
• Polymer Growth	Any two molecular species present can react; slow, random growth takes place.	The growth reaction takes place by the addition of one unit at a time to the active end of the polymer chain.
• Polymer Molecular Weight	Molecular weight rises steadily throughout the reaction. High conversion is required for high molecular weight polymer.	High molecular weight polymer is formed immediately.
• Monomer Concentration During Polymerization	Monomer disappears in the early stages of the polymerization. At an average degree of polymerization of 10, less than 1 weight percent of the monomer remains.	Monomer concentration decreases steadily throughout the reaction.
• Composition of the Polymerization Reaction	A relatively broad calculable distribution of molecular species are present throughout the course of the polymerization.	Mixture contains only monomer, high molecular weight polymer and only about 10-8 part of growing chains. This is true shortly after initiation and at the end of the polymerization (except for the growing chain concentration) since 100% conversion of monomer usually is not achieved.

Polymerization reactions such as that shown for e-aminocaproic acid (A-B monomer) or adipic acid (A-A monomer) with hexamethylenediamine (B-B monomer) can be written in the general forms, as follows:

$$N_0 \text{A-B} \rightarrow \text{A-B}\text{+}[\text{A-B}]_{N_0-2}\text{A-B}$$

$$N_0 \text{A-A} + N_0 \text{B-B} \rightarrow \text{A-A}\text{+}[B-B-A-B]_{N_0-1}\text{B-B}$$

Where N_0 is the number of monomer molecules at the beginning of the reaction. At any given stage of the reaction, if N or 2N functional groups, respectively, have remained unreacted, then there will be N or 2N molecules in the mixture, regardless of size. The total number of functional groups of either type A or B that have reacted is therefore $N_0 - N$.

The extent of a reaction at a given stage of the polymerization is defined by P, which is the ratio of the number of reacted molecules to the initial number of molecules.

$$P = (N_0 - N) / N_0$$
$$N = N_0(1-P)$$

The degree of polymerization $\overline{DP}$ is the original number of molecules divided by the remaining number of molecules and approaches a number equal to the original number of molecules as N approaches one.

$$\overline{DP} = \frac{N_0}{N}$$

Thus, the degree of polymerization can be expressed by:

$$\overline{DP} = \frac{N_0}{N_0(1-P)} = \frac{1}{1-P}$$

Consider a reaction that can be pushed to 90% completion. The number of times an organic chemist realizes this yield in the laboratory is not very often. Yet polymer chemists must obtain yields much better than this to produce high molecular weight step-growth polymers.

$$\overline{DP} = \frac{1}{1-0.9} = 10$$

A decamer is not a high enough molecular weight material to have useful mechanical properties. At $\overline{DP} = 50$ many polymers begin to exhibit sufficient mechanical strength to be useful materials. To realize this $\overline{DP}$ a 98% yield reaction is necessary.

$$50 = \frac{1}{1-P}; P = 0.98$$

What is the effect on $\overline{DP}$ of the presence of an impurity in the monomer or a stoichiometric imbalance? For the polymerization of the A-B monomer, if there is a monofunctional impurity, A, present, and N_0 is the number of molecules of AB while N_1 is the number of molecules of A, then,

$$\overline{DP} = \frac{1 + N_1/N_0}{1 - P + N_1/N_0}$$

With 2% of the impurity A, and a 98% conversion,

$$\overline{DP} = \frac{1 + \dfrac{2}{98}}{1 - 098 + \dfrac{2}{98}} = 25$$

In this case, with the same conversion (98%) an impurity (2%) cuts the molecular weight in half. A stoichiometric imbalance has the same effect on the molecular weight as does an impurity.

$$\overline{DP} = \frac{1+r}{2r(1-P)+1-r}$$

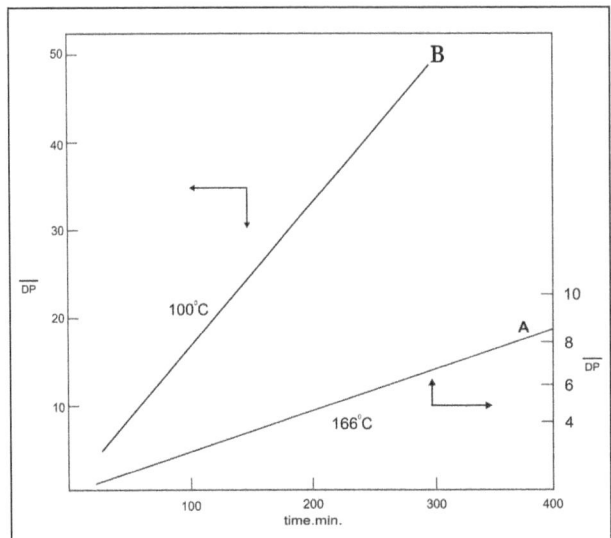

A. Reaction of diethylene glycol with adipic acid. B. Reaction of decamethylene
glycol with adipic acid; added p-toluene sulfonic acid catalyst.

For the polymerization of an A-A monomer with a B-B monomer, if the mole ratio of the two, r, is not one, the expression shows the effect of this imbalance on the molecular weight. For a 98% reaction conversion again, and a 2% excess of B-B (r = 0.98), $\overline{DP} = 33$.

Synthesis of Some Linear Step-growth Polymers

The synthesis of a polyester by the direct reaction of a diacid and a diol generally requires high temperatures, in which case some decarboxylation occurs, resulting in monofunctional acid and a stoichiometric imbalance. The ester interchange reaction in the presence of a catalyst requires much lower temperatures and is the reaction utilized in the synthesis of poly (ethylene terephthalate).

The polymerization takes place with weak base catalysts (calcium acetate, antimony oxide, titanium alkoxide, or titanium oxide) in several stages. The reaction of dimethyl terephthalate with an excess of ethylene glycol is first carried out to obtain monomeric and oligomeric glycol esters. The reaction is carried out at 200 °C, near the boiling point of ethylene glycol.

$$CH_3O-\overset{\overset{O}{\|}}{C}-\underset{}{\bigcirc}-\overset{\overset{O}{\|}}{C}OCH_3 \ + \ 2HOCH_2CH_2OH \xrightarrow[base]{200\ ^\circ C}$$

$$HOCH_2CH_2O-\overset{\overset{O}{\|}}{C}-\bigcirc-\overset{\overset{O}{\|}}{C}-O-CH_2CH_2OH \ +$$

$$HOCH_2-CH_2-O\left[\overset{\overset{O}{\|}}{C}-\bigcirc-\overset{\overset{O}{\|}}{C}-O-CH_2-O\right]_n H$$

$$n \ = \ 2, 3, 4$$

Once all of the methanol has been evolved, the temperature is raised in stages (e.g., 272 °C, 20 min.; 283 °C, 3 hr) under reduced pressure to interchange and remove the over-balance of ethylene glycol. This is a self-balancing reaction, and it is an example in which perfect stoichiometry is not necessary at the outset. The final temperature is higher than the boiling point of ethylene glycol and the melting temperature of the polymer (Tm = 260 °C).

$$\longrightarrow \left[CH_2-CH_2-O-\underset{\substack{\|\\O}}{C}-\langle C_6H_4 \rangle-\underset{\substack{\|\\O}}{C}-O \right] + HOCH_2CH_2OH$$

Fiber and film are manufactured from this polyester. The polymer is melt spun near 270 °C to give a fiber that has high tensile strength. Fabrics, from polyethylene terephthalate) fiber, particularly polyester-cotton blends, are widely used in clothing. The fiber is also used in tire cord.

One method of polycarbonate synthesis is by the reaction of the simplest diacid chloride, phosgene, with bisphenol A in the presence of a base. The reaction can be run either in the presence of pyridine or in a two-phase reaction mixture of methylene chloride and aqueous sodium hydroxide.

$$n\ HO-\langle C_6H_4 \rangle-\underset{\substack{CH_3\\|\\|\\CH_3}}{C}-\langle C_6H_4 \rangle-OH + n\ COCl_2 \xrightarrow[20-30\ °C]{Base}$$

$$\left[\langle C_6H_4 \rangle-\underset{\substack{CH_3\\|\\|\\CH_3}}{C}-\langle C_6H_4 \rangle-O-\underset{\substack{\|\\O}}{C}-O \right]_n + 2n\ HCl \cdot Base$$

The polymer has a low degree of crystallinity (20-40%; Tm = 270 °C, Tg = 145-150 °C), and a very high impact strength. It is a molding resin that has excellent electrical resistance. Polyamides can be prepared by the reaction pf an aliphatic diamine with ft diacid. A perfect stoichiometric balance of hexamethylene diamine and adipic acid is achieved through the formation of a 1:1 salt in methanol. The salt is charged as 60-80% slurry in water and is then heated to purge the air by the steam. The reaction equilibrium favors amidation even in the presence of large amounts of water, so that the initial stages of the polymerization are carried out at 220 °C, and 200-250 psi to achieve an 80-90% conversion after 1 hr. Heating to 270-280 °C while bleeding the steam produces the final material which is extruded in the molten state.

$$\begin{array}{c} ^{\oplus}NH_3(CH_2)_6{}^{\oplus}NH_3 \\ ^{\ominus}OC(CH_2)_4-C-O^{\ominus} \\ \underset{O}{\|}\quad\quad\underset{O}{\|} \end{array} \longrightarrow \left[\underset{}{\overset{H}{N}}-\underset{\substack{\|\\O}}{C}(CH_2)_4\underset{\substack{\|\\O}}{C}-\overset{H}{N}(CH_2)_6 \right]_n$$

The 66 nylon thus obtained is highly crystalline (Tm = 250 °C). It can be spun into fiber having high tensile strength or used as a molding resin that has good impact strength, toughness, and abrasion resistance. The fiber, which can be cold-drawn to increase its orientation and improve its tensile strength, is widely used in the textile industry and in tire cord.

Polyamides composed of aromatic diamines and diacids cannot be obtained directly from the free acid and amine since the aromatic amines are too weakly basic. The reaction of the diamine with a diacid chloride, however, produces high molecular weight polymer. The reaction of isophthaloyl chloride with m-phenylenediamine takes place in a polar solvent such as dimethylacetamide in the presence of a base such as calcium oxide to give a polyaramide that has a polymer melt temperature above 400 °C. This aromatic polyamide remains soluble in the polymerization medium with the aid of the calcium chloride (generated in the reaction) which serves to break up intermolecular hydrogen bonding of the amide functions.

Fiber, wet spun directly from the polymerization medium (and then washed to remove the calcium chloride), is woven into a fabric that has good resistance to chemical attack and is stable to moderately high temperatures. The fiber also is chopped into staple, and paper obtained therefrom is used as a high temperature electrical insulation material.

Polyurethanes are obtained by the reaction of a diisocyanate with a diol. One of the most useful types of polyurethane is that obtained from a moderate molecular weight polymer having alcohol end groups. A polyester with alcohol end groups and the desired molecular weight can be obtained, for example, by the reaction of ethylene glycol with succinic anhydride in the appropriate stoichiometric balance.

The subsequent reaction of the polyester (abbreviated as HOROH) with a diisocyanate gives a polyurethane.

Added to this reaction are catalysts such as stannous octanoate and a tertiary amine. If water is added to this reaction mixture, then carbon dioxide is evolved, and a urethane form is generated by the decarboxylation of the intermediate carbamic acid. In order to generate a uniform cell size in the foam, a surfactant (silicone oil) is also added.

$$\text{RHCO} + \text{H}_2\text{O} \longrightarrow \left[R-\underset{H}{N}-\overset{O}{\underset{||}{C}}-OH \right] \longrightarrow \text{RNH}_2 + \text{CO}_2$$

The amine ends generated through the action of the isocyanate with water may now react with a diisocyanate to afford a urea link. Thus, if the stoichiometric balance of diol, isocyanate and water is adjusted, a high molecular weight polyurethane foam can be obtained.

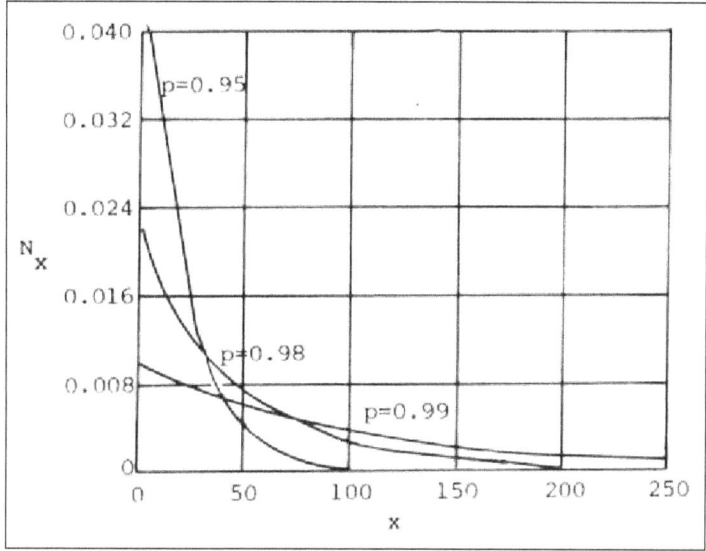

Number or mole fraction distribution in a linear step-growth polymerization for three extents of reaction p.

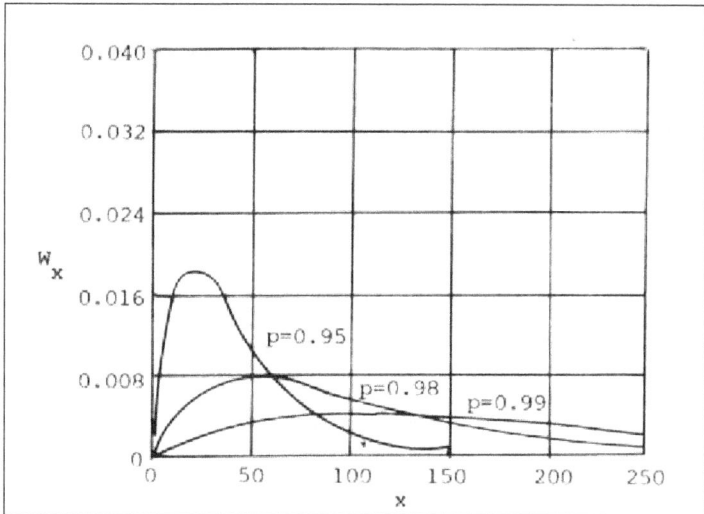

Weight fraction distribution in a linear step-growth polymerization for three extents of reaction p.

Either flexible or rigid foams are realized, depending on the flexibility and molecular weight of the polyester.

These are examples of just a few of the step-growth polymerization reactions carried out industrially, but they illustrate the four major reaction types listed earlier.

Molecular Weight Distribution of Linear Step-growth Polymers

The synthesis of polymers by a step-growth polymerization reaction leads to broad molecular weight distributions. The molecular weight distribution that will result can be calculated by considering the probability, P, that a functional group will have reacted at a given time t. The probability of finding an unreacted functional group then will be $(1 - P)$. P is also the extent of the reaction.

In order to find the probability that a given molecule, selected at random from a polymer sample will contain exactly x structural units, the following is considered, taking as an example the polymerization of an AB monomer, r-aminocaproic acid.

1. The polymer will have $(x - I)$ reacted functional groups and one unreacted end group.

2. The probability of finding a single, reacted group is P, and thus the probability of finding $x - 1$ reacted groups in the same molecule is P^{x-1}.

3. The probability of finding a single unreacted group is $(1 - P)$.

Thus, the probability of finding an x-mer and hence the fraction of the x-mer, is the product of 2 and 3, $P^{x-1}(1 - P)$. If there are N molecules present in the polymer sample, then the fraction of the x -mer (the fraction of molecules that are x-mers) is given by:

$$N_x = NP^{x-1}(1-P)$$

If there were N_0 molecules (monomer) initially, since $N = N_0(1 - P)$, then:

$$N_x = N_0(1-P)^2 P^{x-1}.$$

The molecular weight distribution according to the number fraction at three different extents of reaction are shown in figure. Note that at 99% reaction, monomer is still present in the largest numbers of any of the other species. This is a little misleading, however, since monomer accounts for only a small weight percent of the mixture. If the number of x-mers is replaced by the molecular weight of the x-mers then a weight fraction distribution is obtained:

$$W_x = X N_x / N_0 = X(1-P)^2 P^{x-1}.$$

These distributions have been confirmed experimentally; the agreement is excellent in a number of the polymerizations.

Three-dimensional Network Step-growth Polymers

One of the first synthetic polymers, Bakelite, is a crosslinked network formed by the reaction of phenol (a trifunctional monomer) with formaldehyde (a difunctional monomer). This polymer has the characteristics of a thermosetting resin in that it "sets up" during heat processing, giving a hard, insoluble, higher molecular weight, three-dimensional network instead of a polymer melt. A glyptal resin obtained from the reaction of glycerin with phthaiic anhydride is an example of a polyester resin. A brittle material is obtained from these monomers, so in practice some longer chain diols are introduced.

Such a polymerization will create a gel or an insoluble (but swellable) material at a certain stage of the polymerization where the extent of the reaction may be relatively low, but the average $\overline{DP}$ becomes infinite. The extent of the reaction necessary to reach gellation can be calculated by:

$$P = \frac{2}{f_{av}} - \frac{2}{\overline{DP}f_{av}}$$

Where again P is the extent of the reaction and fav is the average functionality. Gellation occurs when DP becomes infinitely large, so the equation reduces to:

$$P_c = \frac{2}{f_{av}}.$$

The critical extent of polymerization, P_c, at which gellation occurs for the 3:2 ratio of phthaiic anhydride to glycerin (12 functional groups per five monomer molecules; f_{av} - 12/5) is predicted to be at 83.3%.

Another approach to predicting the gel point is through consideration of the probability that a given functional group on a branch unit (i.e. a unit of f > 2) is connected to another branch unit. If again the glyptal type polymerization is considered in which glycerin is abbreviated A_f(f = 3) a diol as A-A, and phthaiic anhydride as B-B, a segment of the polymer can be written as:

$$A_{f-1}\text{-}A\text{-}[\text{-}B\text{-}B\text{-}A\text{-}A]_i\text{-}B\text{-}B\text{-}A\text{-}A\text{-}_{f-1}$$

The criterion for gel formation is that at least one of the f — 1 segments radiating from the end of a segment of the type shown is in turn connected to another branch unit. The probability, α, of this occurring is 1 in / — 1. Thus:

$$a_c = \frac{1}{f-1}.$$

The probability is related to the extent of the reaction in the following way. For the polymer segment shown, if the extents of the reactions for A (alcohol) or B (carboxylic acid) functions are P_A and P_B, respectively, and the ratio of A groups in the branch units to all the A groups is ρ, then the probability that a B group has reacted with a branch A unit is P_B. Likewise, the probability that B has reacted with a non-branch A unit (bifunctional A-A) is P_B $(1-\rho)$. Thus, the probability that a segment of the type shown is formed is:

$$P_A[P_B(1-\rho)P_A]^i P_B\rho$$

Summing over all values of i:

$$\alpha = \frac{P_A P_B \rho}{1 - P_A P_B (1 - \rho)},$$

If $r = N_A / N_B$, then $P_B = r P_A$ and the equation becomes:

$$\alpha = \frac{r P_A^2 \rho}{1 - r P A^2 (1 - \rho)} = \frac{P_B^2 \rho}{r - P_B^2 (1 - \rho)},$$

when an equal number of A and B groups are charged to the polymerization $r = 1 (P = P_A = P_B)$, and the equation simplifies to:

$$\alpha = \frac{P^2 \rho}{1 P^2 (1 - \rho)},$$

when there are no A-A units, as in the case of the polymerization of phthalic anhydride with glycerin, then $\rho = 1$, and:

$$\alpha = r P_A^2 = \frac{P_B^2}{r},$$

when both these conditions apply ($r = 1$ and $\rho = 1$), $\alpha = P^2$.

Thus, for the glyptal resin in which the ratio of alcohol functions to acid functions is 1:1, ($r = 1$, $p = 1$), the critical branching coefficient is $a_c = 1/2$; (1/f— 1). Thus, P = (1/2)$^{1/2}$ = 0.707. Gelation is calculated to occur at a conversion 70.7%. In practice, gelation occurs at 79.5%. This deviation from theory can be ascribed to the fact that the secondary alcohol function on glycerin is less reactive than the primary alcohol functions.

Kinetics of Step Growth Polymerization

The assumption that the reactivity of a functional group on the growing end of the polymer chain is the same as that on a small molecule generally has been observed experimentally. In two cases, when the mobility of the polymer becomes extremely low—as occurs, for example, in a built polymerization in which the medium becomes extremely viscous, especially toward the end of the reaction—and when the mobility is relatively low while the reaction rate constant is high, the reaction slows down.

If the reaction of a diol with a diacid to form polyester is considered, with no acid catalyst added, the diacid acts as its own catalyst, and third order kinetics are observed:

$$\frac{-d[CO_2H]}{dt} = k[CO_2H]^2[OH]$$

[CO_2H] and [OH] are the concentrations of the carboxylic acid and alcohol functions, respectively. If good monomer balance is achieved, then [CO_2H] = [OH]. If these concentrations are expressed in terms of C, then $-dC/dt = kC^3$ and integration gives $2kt = (1/C^2) -$ const. To express C in terms of the extent of the reaction, P, $C = Co\,(1-P)$ and $2kt = 1/[C_0(1-P)]^2 -$ const., where C_0 is the initial concentration of carboxyl or alcohol functions. Thus, $2C_0^2kt = 1/(1-P)^2 -$ const., and a plot of $1/(1-P)^2$ versus t yields a straight line. Note that in the polymerization of diethyleneglycol with adipic acid at 166 °C and no added catalyst, a $\overline{DP}$ of only eight is achieved after six and a half hours.

If, however, a strong acid catalyst such as p-toluenesulfonic acid is added to the polymerization reaction, the rate equation is $-dC/dt = kC^2$ and integration gives $1/C = k't -$ const. Using the relationship $C = C_0(1-P)$, $1/C_0(1-P) = kt -$ const., or $C_0k't = 1/(1-P)$. The second order reaction of a glycol with adipic acid at 109 °C with 0.01 percent of p-toluenesulfonic acid (based on carboxyl functions) gives a polymer of $\overline{DP}$ 48 ($\overline{Mn} \sim 13,500$) after five hours. Since the cost of running a polymerization is very much dependent on the time it takes, many step reaction polymerizations utilize catalysts.

Molecular Weight Distribution in Linear Step-growth Polymers

The molecular weight distribution of a linear condensation or addition polymer can be easily calculated if we assume that each functional group has an equal chance of reacting with other groups regardless of the size of the oligomer. This assumption is called *Flory's equal reactivity principle.* According to this principle, the probability that a given reactive group has reacted is equal to the fraction *p* of all condensed functional groups of the same type, which is called the extend of the reaction. An oligomer containing x repeat units must have undergone x-1 reactions. The probability that this number of reactions has occured is simply the product of all reaction probabilities, i.e. p^{x-1}, whereas the probability of finding an unreacted end group is is 1- p.[1] Hence the total probability, P_x that a given oligomer is composed of exactly x unitis is given by:

$$P_x = (1-p)\,p^{x-1}$$

P_x is equal to the mole fraction, n_x, of x-mers in the reaction mixture of the extend p:

$$P_x = n_x = N_x\,/\,N$$

Then the total number of x-mers is given by:

$$N_x = N\,(1-p)\,p^{x-1}$$

where N is the total number of molecules of all sizes. This number is related to the initial number of monomers or total number of units, N_0, by:

$$N = N_0\,(1-p)$$

where $DP = N_0\,/\,N$ equals the average *degree of polymerization:*

$$DP = N_0\,/\,N = 1\,/\,(1-p)$$

with this substituition, the total number of x-mers can be written in the form:

$$N_x = N_o (1-p)^2 p^{x-1}$$

If the weight of the (condensing) end-groups of each molecule is neglected (for example H + OH), the molecule weight of each molecule is directly proportional to the length of the chain x. Hence, the weight fraction w_x can be written as:

$$w_x = x N_x / N_o$$

The error for condensation polymers will be only significant for low molecular weight polymers.

Then the most probably weight fraction distribution (i.e. the weight-average molecular weight distribution) is given by:

$$w_x = x (1-p)^2 p^{x-1}$$

This distribution is sometimes called Flory-Schulz distribution. The form of this distribution implies that shorter polymer chains are favored over longer ones. It also implies that the length distribution broadens and shifts to higher molecular weight with increasing extend of reaction.

Mole Fraction Distribution of Linear Step-growth Polymers

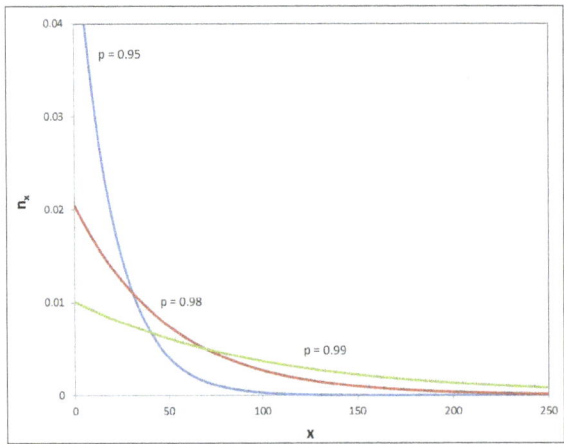

Weight Fraction Distribution of Linear Step-growth Polymers

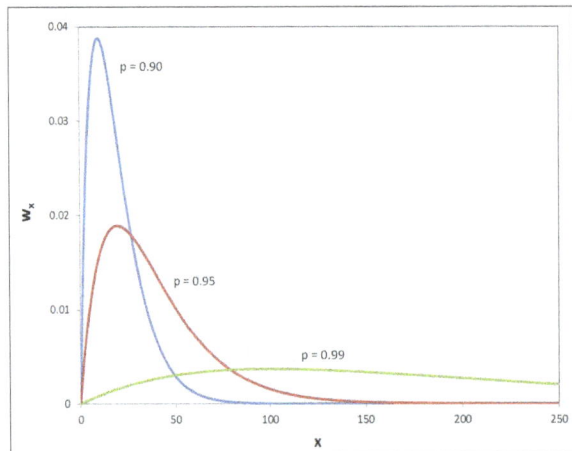

From the expressions above, the number average molecular weight M_n can be easily calculated:

$$M_n = m\,N_o\,/\,N = m\,/\,(1\text{-}p)$$

Where m is the molecular weight of a mer unit. The weight-average molecular weight M_w can be calculated as follows:

$$M_w = \sum_x w_x M_x = m\,(1\text{-}p)^2 \sum_x x^2\,p^{x\text{-}1} = m\,(1+p)\,/\,(1\text{-}p) = (1+p)\,M_n$$

Where $M_x = x{\cdot}m$ is the molecular weight of an x-mer. The ratio of weight and number average of the polymer molecular weight, $M_w\,/\,M_n$, is called the polydispersity or heterogeneity index (D). It is a measure for the broadness of a molecular weight distribution. The polydispersity is unity if all polymer moelcules are of the same size. The polydispersity for the most probable molecular weight distribution is given by:

$$D = m\,(1 + p)\,(1\text{-}p)\,/\,[m(1\text{-}p)] = (1+p)$$

Thus, when the probabilty p is equal to 1 (i.e. all A and B units have reacted at least with one other unit (A-B), the polydispersibility index for the most probable distribution for a linear step-growth reaction approaches 2.

Chain Growth Polymerization

Chain-growth polymerization or chain polymerization is a polymerization technique where unsaturated monomer molecules add onto the active site of a growing polymer chain one at a time. Growth of the polymer occurs only at one (or possibly more) ends. Addition of each monomer unit regenerates the active site.

An example of chain-growth polymerization by ring opening to polycaprolactone.

Polyethylene, polypropylene, and polyvinyl chloride (PVC) are common types of plastics made by chain-growth polymerization. They are the primary component of four of the plastics specifically labeled with recycling codes and are used extensively in packaging.

Mechanism

Chain-growth polymerization can be understood with the chemical equation:

$$(-M-)_n\,(polymer) + M\,(monomer) \rightarrow (-M-)_{n+1}$$

where n is the degree of polymerization and M is some form of unsaturated compound: an alkene (vinyl polymers) or alicyclic compound (ring-opening polymerization) containing molecule.

This type of polymerization result in high molecular weight polymer being formed at low conversion. This final weight is determined by the rate of propagation compared to the rate of individual chain termination, which includes both chain transfer and chain termination steps. Above a certain ceiling temperature, no polymerization occurs.

Steps

Chain-growth polymerization usually has the following steps:

- Chain initiation, usually by means of an initiator which starts the chemical process. Typical initiators include any organic compound with a labile group: e.g. Azo (-N=N-), disulfide (-S-S-), or peroxide (-O-O-). Two examples are benzoyl peroxide and AIBN.

- Chain propagation.

- Chain transfer, terminates the chain, but the active site is transferred to a new chain. This can occur with the solvent, monomer, or other polymer. This process increases the branching of the resulting polymer.

- Chain termination, which occurs either by combination or disproportionation. Termination, in radical polymerization, is when the free radicals combine and is the end of the polymerization process.

The active center can be one of a number of different types:

- Free radical in radical polymerization, for example, polystyrene, sometimes seen as packing peanuts, is produced by polymerizing styrene with Benzoyl peroxide as its radical initiator.

- Carbocation in cationic polymerization, an example is Isobutyl synthetic rubber, initiated by Aluminium chloride ionizing isobutylene.

- Carbanion in anionic polymerization.

- Organometallic complex in coordination polymerization.

Under the necessary reaction conditions, an addition polymerization can be considered a living polymerization. This is most often seen with anionic polymerization as it can be easy to perform without termination steps.

Comparison with other Polymerization Methods

The distinction between step-growth polymerization and chain-growth polymerization was introduced by Paul Flory in 1953, and refers to the difference in reaction mechanisms with step-growth using the functional groups of the monomer compared to the free-radical or ion groups used in chain-growth polymerization.

Chain growth polymerization and addition polymerization (also called polyaddition) are two different concepts. In fact polyurethane polymerizes with addition polymerization (because its polymerization does not produce any small molecules, called "condensate"), but its reaction mechanism is a step-growth polymerization.

Ionic Polymerization

Free radicals are indiscriminate in the compounds they attack, and their non-selective nature in polymerization reactions leads to problems such as chain branching and transfer which affect the structure of the polymer produced. Anionic polymerization overcomes many of these problems.

A typical commercial anionic reaction is the polymerization of styrene using butyllithium, C_4H_9Li, in an inert solvent such as n-hexane. Termination does not occur by polymer-polymer interaction but by reaction with small molecules such as water:

$$(\text{polymer}) - Li + H_2O \rightarrow (\text{polymer}) - H + LiOH$$

This type of polymerization gives rise to very sharp molecular mass distributions because transfer processes are absent. If the solvent is extremely pure, the polymer chains will still be active after all the monomer has been consumed. Such activated systems are known as living polymers and it is possible to continue feeding monomer into the reaction vessel without killing the living chains. The degree of polymerization is simply:

$$n = \frac{[M]}{[I]}$$

Since the chain ends are relatively few in number only a very small amount of water need be present to kill the polymer, and so all ingredients must be rigorously purified. Paradoxically, it is easier to conduct the reaction on an industrial scale than in a laboratory flask. An important source of contamination is the sides of the reaction vessel itself. Since the surface area of a sphere (assuming a spherical reaction vessel of radius r) increases as r^2 while the volume increases as r^3 the problem of surface contamination will be less serious in large, industrial reactors than in laboratory-scale reaction vessels.

Just as negatively charged initiators can be used to start polymerization, so positively charged species can initiate chain growth. The most important commercial operation is the polymerization of iso-butylene giving butyl rubber using aluminium chloride. The reaction conditions are unusual in that high molecular mass polymer is formed very rapidly at very low temperatures (−100 °C for example).

Anionic Polymerization

Anionic polymerization is a form of chain-growth polymerization that encompasses the polymerization of vinyl monomers with strong electronegative groups. This type of polymerization is often used to produce synthetic polydiene rubbers, solution styrene-butadiene rubbers (SBR), and thermoplastic styrenic elastomers.

All monomers with (strong) electronegative substituents polymerize readily in the presence of carbanions. Some electron-withdrawing substituents that stabilize the negative charge through charge delocalization and hence permit anionic polymerization include -CN, -COOR, $-C_6H_5$, and $-CH=CH_2$, to name only a few. Therefore, monomers such as styrenes, dienes, acrylates and methacrylates, aldehydes, epoxides, acrylonitriles and cyanoacrylates readily undergo anionic polymerization reactions.

The electron donors (or initiators) are either electron transfer agents or strong anions. The transfer of an electron from a donor molecule to the vinyl monomer leads to the formation of an anion radical, the so-called carbanion:

$$R-M \quad + \quad H_2C=\underset{X_2}{\overset{X_1}{\underset{|}{\overset{|}{C}}}} \quad \longrightarrow \quad R-CH_2-\underset{X_2}{\overset{X_1}{\underset{|}{\overset{|}{C^-}}}} \quad + \quad M^+$$

Typical electron donors (Lewis bases or nucleophiles) are alkali metals, such as lithium or sodium. Other strong nucleophilic initiators include covalent or ionic metal amides, alkoxides, hydroxides, amines, phosphines, cyanides, and organometallic compounds such as alkyl lithium compounds and Grignard reagents. The initiation proceeds by addition of a neutral (B:) or negative (B:$^-$) nucleophile to the monomer.

The kinetics of an anionic polymerization consists of initiation, polymerization and termination. For example, the initiation and polymerization of styrene with potassium amide proceeds as follows:

$$KNH_2 \Leftrightarrow K^+ + NH_2^-$$

$$NH_2^- + M \rightarrow NH_2M^-$$

$$NH_2M_n^- + M \rightarrow NH_2M_{n+1}^-$$

$$NH_2M_n\text{-} + NH_3 \rightarrow NH_2M_nH + NH_2^-$$

The "Gegen" ion, K^+, can be omitted from the scheme above, because it is dissolved ("free") in a media of comparatively high dielectric constant.

In carefully controlled systems (pure reactants and inert solvents), an anionic polymerization does not undergo termination reactions. Hence, the chains will remain active indefinitely unless there is deliberate termination or chain transfer. This has two important consequences:

1. The number average molecular weight, M_n, of the polymer can be calculated from the amount of initiator and amount of consumed monomer, because the degree of polymerization is the ratio of the moles of monomer consumed to the moles of the initiator added: M_n = MW_0 $[M_o]$ / $[I]$, where MW_0 is the molecular weight of the repeat unit and $[M_o]$ and $[I]$ the (initial) concentrations of the monomer and the initiator.

2. Since all chains are initiated at roughly the same time, the polymer synthesis can be done in a controlled manner. In fact, it is the only one that leads to well defined and nearly mono-disperse molecular weight distribution (Poisson distribution) and structural and compositional uniformity.

This type of polymerization is called living polymerization.

Anionic polymerization can also be used to functionalize polymers. The end-groups are usually added at the end of the polymerization. End-groups that have been used in the functionalization include -OH, -SH, -NH$_2$, COCH$_3$, -COOH, and epoxides, to name only a few.

Free Radical Polymerization

All the monomers from which addition polymers are made are alkenes or functionally substituted alkenes. The most common and thermodynamically favored chemical transformations of alkenes are addition reactions. Many of these addition reactions are known to proceed in a stepwise fashion by way of reactive intermediates, and this is the mechanism followed by most polymerizations. A general diagram illustrating this assembly of linear macromolecules, which supports the name chain growth polymers, is presented here. Since a pi-bond in the monomer is converted to a sigma-bond in the polymer, the polymerization reaction is usually exothermic by 8 to 20 kcal/mol. Indeed, cases of explosively uncontrolled polymerizations have been reported.

It is useful to distinguish four polymerization procedures fitting this general description:

- Radical Polymerization: The initiator is a radical and the propagating site of reactivity (*) is a carbon radical.

- Cationic Polymerization: The initiator is an acid and the propagating site of reactivity (*) is a carbocation.

- Anionic Polymerization: The initiator is a nucleophile and the propagating site of reactivity (*) is a carbanion.

- Coordination Catalytic Polymerization: The initiator is a transition metal complex and the propagating site of reactivity (*) is a terminal catalytic complex.

Radical Chain-growth Polymerization

Virtually all of the monomers described above are subject to radical polymerization. Since this can be initiated by traces of oxygen or other minor impurities, pure samples of these compounds are often "stabilized" by small amounts of radical inhibitors to avoid unwanted reaction. When radical polymerization is desired, it must be started by using a radical initiator, such as peroxide or certain azo compounds. The formulas of some common initiators and equations showing the formation of radical species from these initiators are presented below.

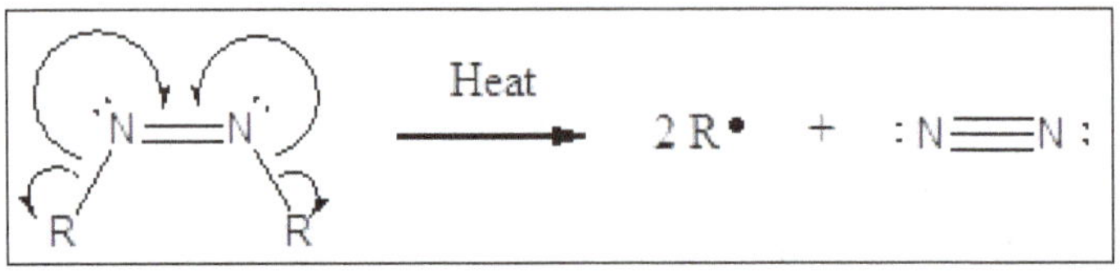

By using small amounts of initiators, a wide variety of monomers can be polymerized. One example of this radical polymerization is the conversion of styrene to polystyrene, shown in the following diagram. The first two equations illustrate the initiation process, and the last two equations are examples of chain propagation. Each monomer unit adds to the growing chain in a manner that generates the most stable radical. Since carbon radicals are stabilized by substituents of many kinds, the preference for head-to-tail region selectivity in most addition polymerizations is understandable. Because radicals are tolerant of many functional groups and solvents (including water), radical polymerizations are widely used in the chemical industry.

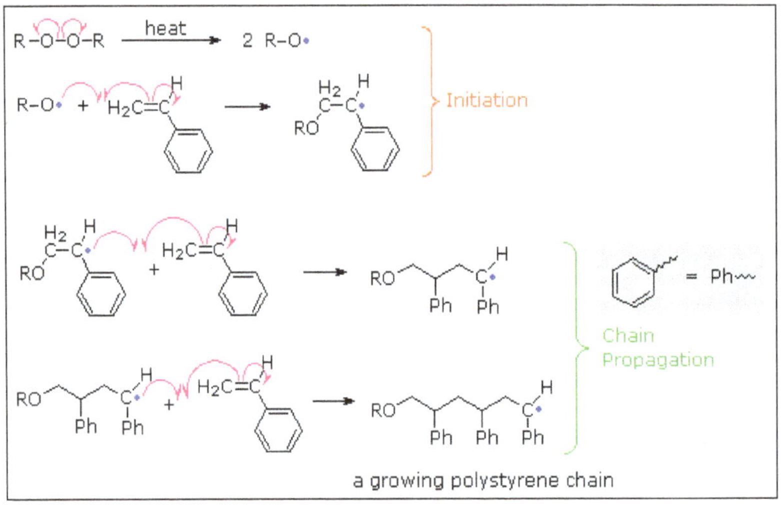

a growing polystyrene chain

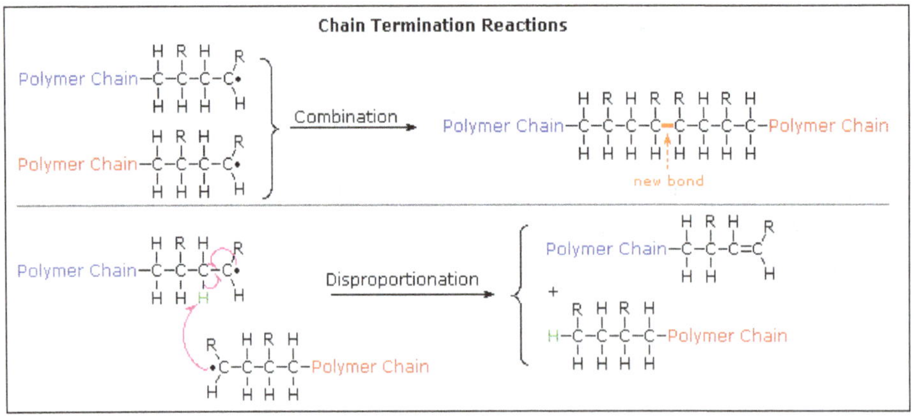

Chain Termination Reactions

In principle, once started a radical polymerization might be expected to continue unchecked, producing a few extremely long chain polymers. In practice, larger numbers of moderately sized chains are formed, indicating that chain-terminating reactions must be taking place. The most common termination processes are Radical Combination and Disproportionation. These reactions are illustrated by the following equations. The growing polymer chains are colored blue and red, and the hydrogen atom transferred in disproportionation is colored green. Note that in both types of termination two reactive radical sites are removed by simultaneous conversion to stable products. Since the concentration of radical species in a polymerization reaction is small relative to other reactants (e.g. monomers, solvents and terminated chains), the rate at which these radical-radical termination reactions occurs is very small, and most growing chains achieve moderate length before termination.

The relative importance of these terminations varies with the nature of the monomer undergoing polymerization. For acrylonitrile and styrene combination is the major process. However, methyl methacrylate and vinyl acetate are terminated chiefly by disproportionation.

Another reaction that diverts radical chain-growth polymerizations from producing linear macromolecules is called chain transfer. As the name implies, this reaction moves a carbon radical from one location to another by an intermolecular or intramolecular hydrogen atom transfer (colored green). These possibilities are demonstrated by the following equations:

Chain transfer reactions are especially prevalent in the high pressure radical polymerization of ethylene, which is the method used to make LDPE (low density polyethylene). The 1°-radical at the end of a growing chain is converted to a more stable 2°-radical by hydrogen atom transfer. Further polymerization at the new radical site generates a side chain radical, and this may in turn lead to creation of other side chains by chain transfer reactions. As a result, the morphology of LDPE is an amorphous network of highly branched macromolecules.

Kinetics Model of Bulk Free Radical Polymerization

Free-radical bulk polymerization is a versatile process because it can be carried out on many monomers and at a wide range of temperatures. It is well-known that the free-radical bulk polymerization of vinyl monomers (derivatives of acrylic and methacrylic acids, vinyl acetate, styrene, ethylene, and so on) is characterized by the autoacceleration phenomenon. The free-radical polymerization of these monomers can be explained by the classical theory up to a certain monomer conversion. After this critical monomer conversion, the autoacceleration of the polymerization appears. The onset of the autoacceleration is defined as the moment when the polymerization rate departs from the value expected according to the classical theory of free-radical polymerization. The onset and the intensity of the autoacceleration are determined by the type of monomer, type and concentration of initiator, temperature and other reaction conditions. This phenomenon is particularly apparent in the bulk polymerization of methyl methacrylate (MMA) and is highly undesirable in industrial applications, as it may lead to the thermal runaway of the process, thus causing depolymerization and the plugging of equipment. A number of theoretical explanations and kinetic models were developed in order to explain the phenomenon of autoacceleration. The first models considered only the decrease of the chain termination rate constant as a result of an increased viscosity of the reaction system. Later efforts focused on investigating changes of both the chain propagation and chain termination rate constants due to changes in the viscosity. Later still, theories attempted to explain the occurrence of autoacceleration in the rate of polymerization as a result of entanglement of growing macromolecular chains. O'Driscoll introduced the gel effect index as a measure of the severity of the kinetic effect. The model developed by Chiu et al. used the conversion, temperature and weight-average molecular weight to determine the relative influence of the reaction and the diffusion on the rate of polymerization. Free volume theories considered the decrease in volume on disposal for the movement of growing macromolecules. Kargin and Kabanov and Korolev et al. suggested that the autoacceleration phenomenon could be explained by the supramolecular organization of the liquid MMA. Roschupkin et al. found that poly (methyl methacrylate) (PMMA) grains are formed during the polymerization of MMA and suggested that the autoacceleration is a consequence of the growth of the grain surface.

A model written in terms of the moment generating function and in terms of the moments of molecular weight distribution, and completed with relations that quantify the gel and glass effects was successful in describing MMA bulk polymerization. A mathematical model was developed for the batch MMA polymerization reactor system by Rafizadeh. The model includes the complete process, therefore, using the heaters and water valve signals, which make it possible to calculate the process states. Hence, this model is suitable to determine an optimal temperature trajectory during the course of the polymerization and control strategy. More recently, a simple semi-empirical model relating the degree of conversion and the polymerization rate to the time and temperature was developed. Model parameters were calculated from isothermal differential scanning calorimetry (DSC) experiments and then successfully applied to predict monomer conversion in non-isothermal experiments. Sangwai et al. used an empirical model that involves only monomer conversion and temperature, and accounts for the gel and the glass effect to describe the polymerization of MMA in a rheometer-reactor assembly in isothermal and non-isothermal conditions. During the last decade, successful results were achieved through the use of the pulsed laser polymerization technique to determine the values of k_p and k_t for free-radical polymerization. Barner-Kowollik et al. have presented an extensive review of the experimental methods used to study the dependence of kt on the conversion and chain length. Buback et al. have used this technique to obtain the chain length dependence of k_t for MMA polymerization.

The existing theories, however, have not been completely verified experimentally. Their main shortcoming is that they take only the onset of acceleration as a characteristic point on the polymerization rate vs time curve. We have shown that there are some additional characteristic points in the case of methyl-2, ethyl-3, butyl3 dodecylmethacrylates and the polymerization of styrene: the maximum polymerization rate and the two inflection points before and after that maximum.

We have focused on determining the existence of these characteristic points, as well as on testing a mathematical model of the free-radical bulk polymerization developed earlier by our research group and successfully tested on the polymerization of styrene. Experimental data were obtained by using DSC to follow the bulk free-radical polymerization of MMA at different temperatures.

Materials and Methods

MMA monomer was washed two times with 10% sodium hydroxide solution to remove the inhibitor. Then, the MMA was washed two times with distilled water, dried over anhydrous calcium chloride and vacuum distilled. Initiator 2,2'-azobisisobutyronitrile (AIBN, Merck, Darmstadt, Germany) was recrystallized from methanol before usage. A solution of 0.5 wt. % AIBN in MMA was prepared. Approximately 5–10 mg of the solution was placed in a hermetic aluminum DSC pan and sealed with an aluminum lid. The bulk polymerization of MMA was performed in a Q20 DSC (TA Instruments, New Castle, DE, USA) under isothermal conditions at 60, 70, 80 and 90 °C. Every experiment was repeated three times. The temperature and heat flow scales were calibrated using the melting of high-purity indium. Nitrogen was used as a purge gas with a flow rate of 50 cm³ min⁻¹.

The figure below shows the DSC thermograms of MMA free-radical bulk polymerization at different temperatures. Three characteristic moments can clearly be observed in the figure below: the onset of acceleration (point M), the maximum of the polymerization rate (point S) and the end of polymerization (point K).

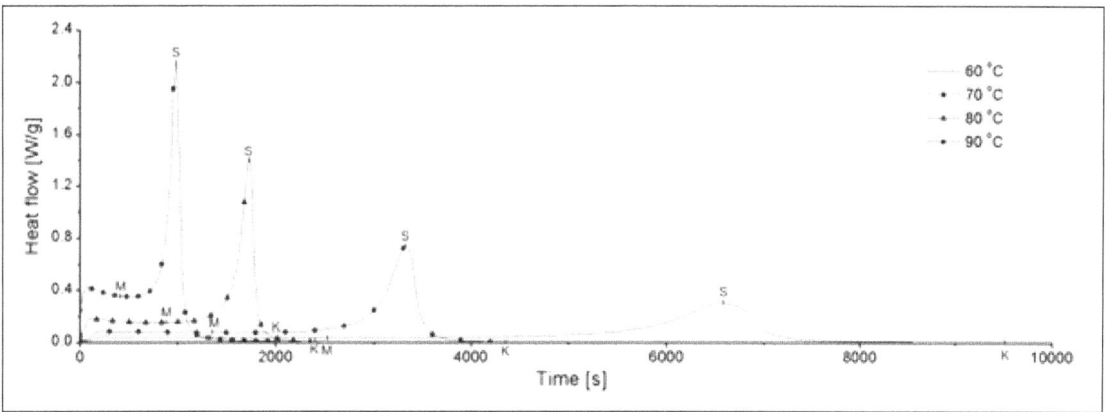

DSC thermograms of MMA free-radical polymerization at different temperatures.

Point M was determined as the minimum and point S as the maximum in the DSC thermogram. Point K was determined as the moment when the isothermal DSC curve becomes horizontal.

To find the monomer conversion degree the following formula was applied:

$$X = \frac{\int_0^\tau (dH/d\tau)d\tau}{\int_0^{\tau K}(dH/d\tau)d\tau + H_D}$$

where X is the monomer conversion, $X = (C_{Mo} - C_M)/C_{Mo}$; C_{Mo} and C_M are the initial monomer concentration and concentration after time τ, respectively; dH is the heat evolved by polymerization during an infinitesimal time (dτ); $^\tau K$ is the time required to achieve point K; HD is the heat evolved during polymerization of an unreacted monomer left after point K as determined by dynamic DSC measurement. The conversion of MMA at different temperatures is shown in figure below.

The conversion vs time curves of the MMA polymerization exhibit portions that have an 'S' shape, characteristic for autoacceleration. The final conversion of MMA has higher values at the higher temperatures.

The rate of polymerization, R_{pol} = dX/dτ, is equal to the slope of the conversion vs time curve. The acceleration, $dR_{pol}/dτ = d^2X/dτ$, is equal to the slope of the R_{pol} vs time curves. The polymerization of MMA exhibits the following characteristic points: the onset of acceleration (M), the maximum acceleration (P) and the maximum polymerization rate (S), proceeded by the deceleration stage with minimum (R) and final conversion (K).

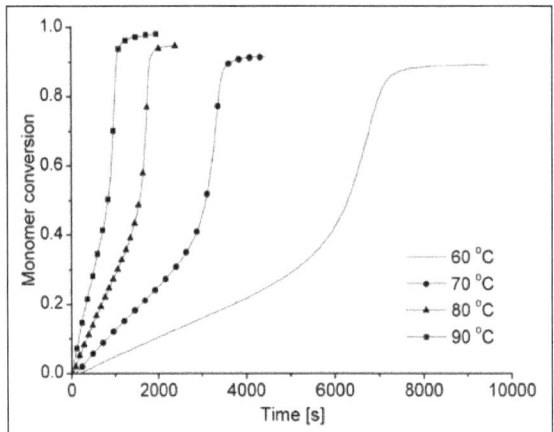

MMA conversion vs time at different temperatures.

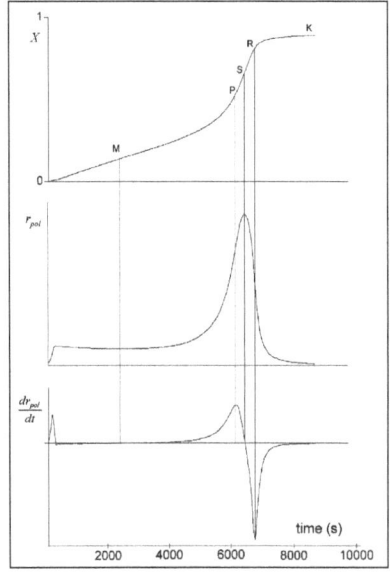

Characteristic points on the conversion (X), rate of polymerization (rpol) and acceleration (drpol/dt) vs. time (t) curves obtained by transformation of the DSC curve of MMA polymerization at 60 °C.

The time and MMA conversions, required to achieve the characteristic points, are given in the table below. The conversion X_M at the onset of the autoacceleration increases as the polymerization temperature increases, in accordance with published data.

Kinetic Model

Three regions can be observed during the polymerization of MMA. The first is up to point M. The second is the acceleration portion from point M to point S. The third is the deceleration from point S to point K. A potential kinetic model should fit the conversion vs time curves, but it should also exhibit the derivatives $(dX/d\tau)$ and $(d^2X/d\tau)$ that resemble the shape of the corresponding curves derived from the experimental results. A potential model should fulfill the following conditions: conversion starts from zero ($\tau = 0$, $X = 0$), the first derivative $(dX/d\tau)$ has a local minimum (M), then reaches the first inflection point (P) followed by a maximum (S) and second inflection point (R), becoming horizontal (K) at the end.

It is generally accepted that the first part of the conversion vs time curve can be explained by the classical free-radical polymerization. Hence, the polymerization rate Rpol in that part can be expressed by the well-known equation:

$$R_{poll} = -\frac{dC_M}{d\tau} = C_{M0}\frac{dX_1}{d\tau} = k_{pol,1}\, C_M C_1^{0.5} = k_{pol} C_{M0}(1-X_1)C_1^{0.5}$$

Table: Time and MMA conversion, necessary to achieve characteristic points (determined from DSC curves).

T (°C)	Point M		Point P		Point S		Point R		Point K	
	τ_M (s)	X_M	τ_P (s)	X_P	τ_S (s)	X_S	τ_R (s)	X_R	τ_R (s)	X_K
60	2520	0.133	6371	0.530	5732	0.688	7068	0.826	9480	0.892
60	2460	0.142	6262	0.535	6571	0.676	6889	0.815	8880	0.892
60	2340	0.123	6466	0.542	6840	0.714	7085	0.819	8882	0.888
70	1530	0.197	3193	0.556	3360	0.679	3633	0.856	4530	0.924
70	1530	0.193	3217	0.599	3330	0.734	3427	0.843	4440	0.915
70	1440	0.181	3249	0.586	3390	0.734	3498	0.844	4980	0.919
80	876	0.247	1709	0.650	1764	0.773	1805	0.858	3192	0.951
80	816	0.238	1688	0.657	1738	0.776	1778	0.859	2438	0.953
80	816	0.238	1688	0.657	1738	0.7476	1778	0.859	2438	0.953
90	542	0.308	963	0.674	992	0.757	1030	0.872	1980	0.978
90	548	0.314	952	0.678	988	0.801	1022	0.877	2010	0.981
90	540	0.306	966	0.695	990	0.786	1017	0.881	2070	0.975

The reaction rate is first order with respect to the monomer and an order of 0.5 with respect to the initiator concentration.

$$R_{poll} = -\frac{dC_M}{d\tau} = C_{M0}\frac{dX_1}{d\tau} = k_{pol,1}\, C_M C_1^{0.5} = k_{pol} C_{M0}(1-X_1)C_1^{0.5}.$$

Here, X_1 is the conversion of a part of the monomers that polymerize according to the classical theory of polymerization and $k_{pol,1}$ is the corresponding polymerization rate constant; C_I is the initiator concentration after time τ. It is usually supposed that there is a negligible change of the initiator concentration during the polymerization, that is, $C_I = C_{I0} = $ const. Hence, the rate of polymerization is described by the equation:

$$\frac{dX_1}{d\tau} = k_1 \cdot (1-X_1)$$

Where $k_1 = k_{pol,1} C_I^{1/2}$ is the rate constant for the first-order reaction. All the theories of the autoacceleration in the rate of polymerization assume that the polymer produced during the autoacceleration stage has a catalytic effect on the polymerization. Hence, we propose that, in that stage, the rate of polymerization depends not only on the concentration of the residual monomer M but also on the amount of the created polymer P, that is, equation:

$$R_{pol2} = C_{M0} \frac{dX_2}{d\tau} = k_{pol,2} C_M C_p C_I^{0.5}$$

This equation can easily be transformed to equation:

$$\frac{dX_2}{d\tau} = k_2 \cdot X_2 \cdot (1 - X_2)$$

where X_2 is the conversion of a portion of the monomers that polymerize according to the autoacceleration and deceleration mechanisms of polymerization and $k_2 = k_{pol,2} C_{Mo} C_I^{1/2}$ is the corresponding polymerization rate constant. It should be noted that equation $\frac{dX_2}{d\tau} = k_2 \cdot X_2 \cdot (1 - X_2)$ is a parabola that has a maximum. Hence, it can be used to describe both the acceleration and deceleration portions of the polymerization rate vs time curves.

After integration and rearrangement, equations $\frac{dX_1}{d\tau} = k_1 \cdot (1 - X_1)$ and $\frac{dX_2}{d\tau} = k_2 \cdot X_2 \cdot (1 - X_2)$ become both equations respectively:

$$X_1(\tau) = 1 - e^{-k_{1r}}$$

$$X_2(\tau) = \frac{1}{1 + e^{-k_2(\tau - \tau_2 \, max)}}$$

where τ_{2max} is the time required to achieve the maximum rate of the autoacceleration stage.

Equation $X_1(\tau) = 1 - e^{-k_{1r}}$ and equation $X_2(\tau) = \frac{1}{1 + e^{-k_2(\tau - \tau_2 \, max)}}$ are conversions according to classical first-order polymerization and acceleration, respectively. At the end of the polymerization, the achieved final conversion X_K consists of two fractions: A is the monomer fraction polymerized by autoacceleration (that is, equation $X_2(\tau) = \frac{1}{1 + e^{-k_2(\tau - \tau_2 \, max)}}$) and ($X_K$ a) is the monomer fraction polymerized by a first-order reaction (that is, equation $X_1(\tau) = 1 - e^{-k_{1r}}$). Based on these assumptions, the dependence of the conversion on time can be presented by a mathematical model. The first addend corresponds to the first-order reaction and the second addend to the autoacceleration reaction. The model takes into account the overlap of these two contributions from the beginning of the polymerization.

Table: The values of the parameters in equation below determined for MMA free-radical polymerization at different temperatures (obtained using the method of least squares).

T (°C)	k_1 (s⁻¹) 10³	k_2 (s⁻¹) 10³	a	τ_{2max} (s)	s.d.	R^a
60	0.344	2.34	0.621	6301	0.027	0.997
60	0.339	2.49	0.600	6202	0.026	0.997
60	0.343	2.45	0.620	6363	0.026	0.997
70	0.661	4.14	0.598	3137	0.028	0.996
70	0.673	5.78	0.588	3136	0.032	0.995
70	0.688	5.23	0.597	3183	0.027	0.997
80	1.27	9.60	0.573	1652	0.028	0.997
80	1.34	10.1	0.574	1619	0.033	0.995
80	1.34	9.84	0.574	1630	0.029	0.996
90	2.26	14.6	0.552	930	0.023	0.998
90	2.30	15.0	0.551	920	0.024	0.998
90	2.34	16.2	0.556	926	0.026	0.997

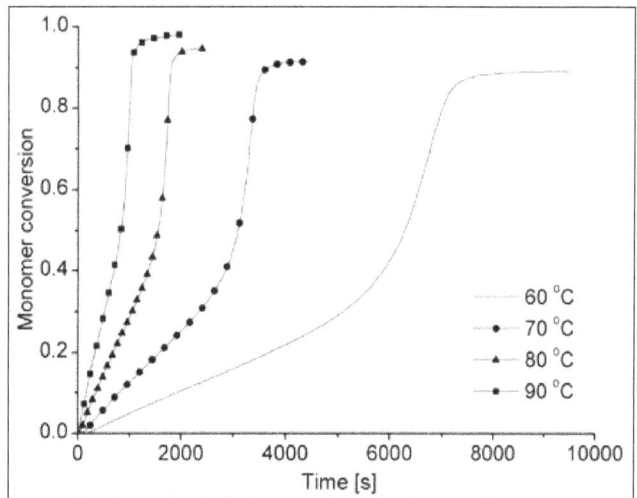

MMA conversion vs time at 60 °C.

This assumption was tested by fitting the model to the experimental data.

$$X(\tau) = (X_K - a) \cdot (1 - e^{-k_1 \tau}) + \frac{a}{1 + e^{-k_2(\tau - \tau_{2max})}}$$

In above equation, the values for X and t were obtained from the experimental DSC data, while k_1, k_2, a and τ_{2max} were calculated using the method of least squares. The values of all calculated parameters are given in Table. The proposed mathematical model describes the experimental dependence of the monomer conversion degree on the polymerization time. The low values of the s.d. and the high correlation coefficients confirm this conclusion. The proposed model includes both addends from the beginning to the end of polymerization. The first addend in equation above gives the main contribution to the value of the conversion, in comparison with the negligible contribution of the second addend. The contribution of the second addend becomes important after a certain amount of time after which it increases and becomes dominant in the second portion of the polymerization.

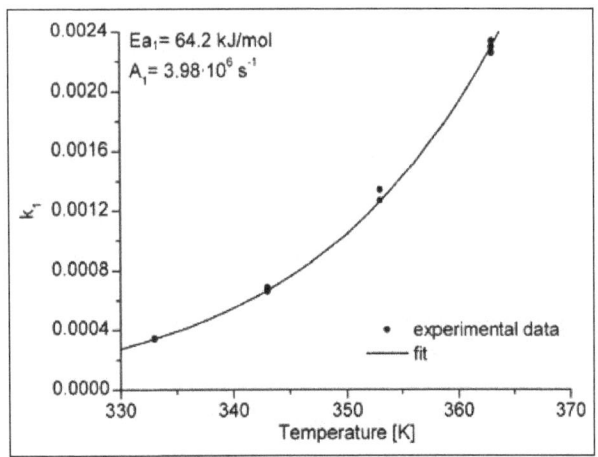

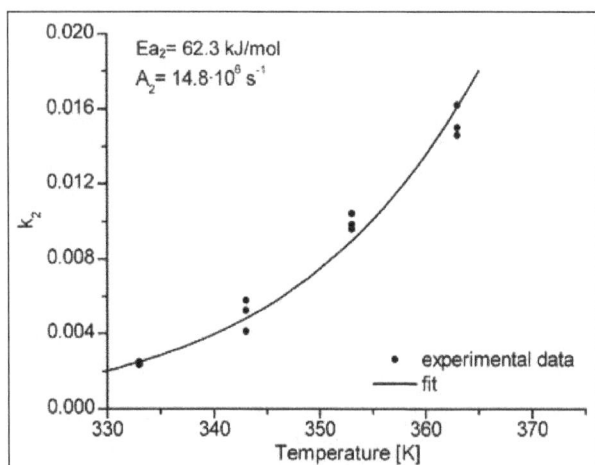

The dependence of k_1 on temperature for MMA polymerization.

The dependence of k_2 on temperature for MMA polymerization.

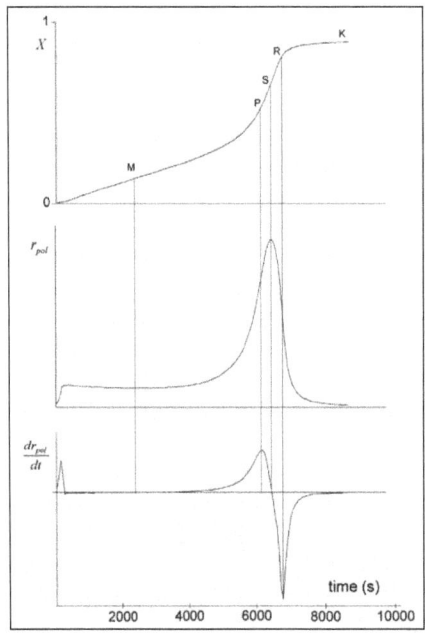

Conversion (X), rate of polymerization (dX/dτ) and acceleration (d²X/dτ) vs. time (τ) curves obtained by our mathematical model (equation 6 using parameters from Table, describing MMA polymerization at 60 °C).

The values of k_2 are seven to eightfold higher than the values of k_1. This means that monomers that react by autoacceleration react much faster. The apparent activation energies and pre-exponential factors for the two reaction rate constants were determined according to the Arrhenius law. The values of the two apparent activation energies are very close but the pre-exponential factor of autoacceleration is approximately fourfold higher than the first-order reaction.

The conversion (X), rate of polymerization (dX/dτ) and acceleration (d²X/dτ) vs time (τ) curves calculated by our mathematical model have the same trends as the corresponding curves obtained from the experimental data. The model fulfills all the conditions set during model development and mentioned earlier.

The polymerization of MMA exhibits the same characteristic points as the polymerization of other lower alkylmethacrylates and styrene: the onset of acceleration (M), the maximum acceleration (P), and the maximum polymerization rate (S) proceeds by a deceleration stage with a minimum (R) and a final conversion (K). A kinetic model that has been tested earlier on styrene polymerization28 was adjusted and applied to describe MMA free-radical bulk polymerization. The model is composed of two contributions, one from the first-order reaction and the other from the autoacceleration reaction. Experimental data of MMA conversion dependence on reaction time are well described by the proposed kinetic model. The proposed model provides values for the four parameters. Model parameter a is the monomer fraction polymerized by autoacceleration. Parameter τ_{max} corresponds to the time required to achieve the maximum rate of polymerization. Parameter k_1 is a compound constant that incorporates three constants: k_i (initiation rate constant), k_p (propagation rate constant) and k_t (termination rate constant), that is, $k_1 = k_p (k_i \cdot C_I/k_t)^{1/2}$. It is not possible, however, to obtain separate values for these three constants. Parameter k_2 is also a compound constant that describes the polymerization of the monomer fraction by autoacceleration. It can be ascribed to the polymerization of an organized fraction of monomers.

Polycondensation

Polycondensation Reaction

Polycondensation is the process of forming polymers by the combination of different monomers. The process is frequently accompanied by the release of various subsidiary low-molecular products (water, alcohol, salt).

For polycondensation, the following monomers are characteristic: compounds with molecules of at least two functional groups. They are usually divided for convenience into three groups:

- Identical functional groups which do not react among each other (diamines);

- Different functional groups which may react among each other and thus form polymers (amino acids);

- Identical functional groups which may react among each other, forming simple polyethers.

In this process, reactions of functional groups of monomers are sometimes possible not only with other groups, but among each other. This explains why so many polymers can be formed.

General chemical structure of one type of condensation polymer.

Polycondensation is a process of several successive stages. Monomers are used up relatively quickly (at the early stage of reaction). After this, a high-molecular polymer is formed from other oligomers which were previously formed by functional groups. In this process, different exchange reactions are possible. In the polycondensation process, numerous polymers are formed which take part in metabolism and various biological processes in the human body.

Branching

In polymer chemistry, branching occurs by the replacement of a substituent, e.g., a hydrogen atom, on a monomer subunit, by another covalently bonded chain of that polymer; or, in the case of a graft copolymer, by a chain of another type. Branched polymers have more compact and symmetrical molecular conformations, and exhibit intra-heterogeneous dynamical behavior with respect to the unbranched polymers. In crosslinking rubber by vulcanization, short sulfur branches link polyisoprene chains (or a synthetic variant) into a multiply branched thermosetting elastomer. Rubber can also be so completely vulcanized that it becomes a rigid solid, so hard it can be used as the bit in a smoking pipe. Polycarbonate chains can be crosslinked to form the hardest, most impact-resistant thermosetting plastic, used in safety glasses.

Polymers which are branched but not crosslinked are generally thermoplastic. Branching sometimes occurs spontaneously during synthesis of polymers; e.g., by free-radical polymerization of ethylene to form polyethylene. In fact, preventing branching to produce linear polyethylene requires special methods. Because of the way polyamides are formed, nylon would seem to be limited to unbranched, straight chains. But "star" branched nylon can be produced by the condensation of dicarboxylic acids with polyamines having three or more amino groups. Branching also occurs naturally during enzymatically-catalyzed polymerization of glucose to form polysaccharides such as glycogen (animals), and amylopectin, a form of starch (plants). The unbranched form of starch is called amylose.

The ultimate in branching is a completely crosslinked network such as found in Bakelite, a phenol-formaldehyde thermoset resin.

Branch point in a polymer.

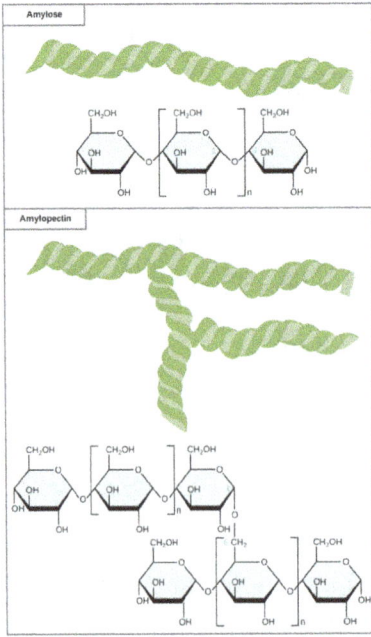

A branched polysaccharide.

Special Types of Branched Polymer

- A graft polymer molecule is a branched polymer molecule in which one or more of the side chains are different, structurally or configurationally, from the main chain.

- A star-shaped polymer molecule is a branched polymer molecule in which a single branch point gives rise to multiple linear chains or arms. If the arms are identical the star polymer molecule is said to be regular. If adjacent arms are composed of different repeating sub-units, the star polymer molecule is said to be variegated.

- A comb polymer molecule consists of a main chain with two or more three-way branch points and linear side chains. If the arms are identical the comb polymer molecule is said to be *regular*.

- A brush polymer molecule consists of a main chain with linear unbranched side chains and where one or more of the branch points has four-way functionality or larger.

- A polymer network is a network in which all polymer chains are interconnected to form a single macroscopic entity by many crosslinks. See for example thermosets or interpenetrating polymer networks.

- A dendrimer is a repetitively branched compound.

Dendrimer synthesis first generation Newkome 1985

Branching in Radical Polymerization

In free radical polymerization, branching occurs when a chain curls back and bonds to an earlier part of the chain. When this curl breaks, it leaves small chains sprouting from the main carbon backbone. Branched carbon chains cannot line up as close to each other as unbranched chains can. This causes less contact between atoms of different chains, and fewer opportunities for induced or permanent dipoles to occur. A low density results from the chains being further apart. Lower melting points and tensile strengths are evident, because the intermolecular bonds are weaker and require less energy to break.

The problem of branching occurs during propagation, when a chain curls back on itself and breaks - leaving irregular chains sprouting from the main carbon backbone. Branching makes the polymers less dense and results in low tensile strength and melting points. Developed by Karl Ziegler and Giulio Natta in the 1950s, Ziegler-Natta catalysts (triethylaluminium in the presence of a metal(IV) chloride) largely solved this problem. Instead of a free radical reaction, the initial ethene monomer inserts between the aluminium atom and one of the ethyl groups in the catalyst. The polymer is then able to grow out from the aluminium atom and results in almost totally unbranched chains. With the new catalysts, the tacticity of the polypropene chain, the alignment of alkyl groups, was also able to be controlled. Different metal chlorides allowed the selective production of each form i.e., syndiotactic, isotactic and atactic polymer chains could be selectively created.

Polymerization of 1,3-butadiene

However there were further complications to be solved. If the Ziegler-Natta catalyst was poisoned or damaged then the chain stopped growing. Also, Ziegler-Natta monomers have to be small, and it was still impossible to control the molecular mass of the polymer chains. Again new catalysts, the metallocenes, were developed to tackle these problems. Due to their structure they have less premature chain termination and branching.

Branching Index

The branching index measures the effect of long-chain branches on the size of a macromolecule in solution. It is defined[6] as $g = <s_b^2>/<s_l^2>$, where s_b is the mean square radius of gyration of the branched macromolecule in a given solvent, and s_l is the mean square radius of gyration of an otherwise identical linear macromolecule in the same solvent at the same temperature. A value greater than 1 indicates an increased radius of gyration due to branching.

References

- Polymerization, science: britannica.com, Retrieved 17 March, 2019
- Calculate-density-polymer-blend-8516878: sciencing.com, Retrieved 20 January, 2019
- Molecular-Weight-Distribution-Stepgrowth, polymer-chemistry: polymerdatabase.com, Retrieved 21 May, 2019
- Introduction-polymers, chemistry, science, science-maths-technology: open.edu, Retrieved 23 August, 2019
- Anionic-polymerization, polymer-chemistry: polymerdatabase.com, Retrieved 30 June, 2019
- Free-Radical-Polymerization, Polymerization, A-Reactions-of-Alkenes, A-Organic-Chemistry: libretexts.org, Retrieved 27 February, 2019
- Kinetic-modeling-of-bulk-free-radical-polymerization-of-methyl-methacrylate: researchgate.net, Retrieved 7 July, 2019
- Polymerization-and-polycondensation-reactions: melscience.com, Retrieved 2 April, 2019

3

Types of Polymer Processing

The manufacturing activity of transforming raw polymeric materials into final products of desired properties and microstructure are known as polymer processing. Some of the types of polymer processing techniques and processes are molding, extrusion, spinning and vulcanization. The topics elaborated in this chapter will help in gaining a better perspective about these types of polymer processing.

Plastics

Plastic is the polymeric material that has the capability of being molded or shaped, usually by the application of heat and pressure. This property of plasticity, often found in combination with other special properties such as low density, low electrical conductivity, transparency, and toughness, allows plastics to be made into a great variety of products. These include tough and lightweight beverage bottles made of polyethylene terephthalate (PET), flexible garden hoses made of polyvinyl chloride (PVC), insulating food containers made of foamed polystyrene, and shatterproof windows made of polymethyl methacrylate.

Most automobile interiors today are made largely of plastic parts.

Composition, Structure and Properties of Plastics

Many of the chemical names of the polymers employed as plastics have become familiar to consumers, although some are better known by their abbreviations or trade names. Thus, polyethylene terephthalate and polyvinyl chloride are commonly referred to as PET and PVC, while foamed polystyrene and polymethyl methacrylate are known by their trademarked names, Styrofoam and Plexiglas (or Perspex).

Industrial fabricators of plastic products tend to think of plastics as either "commodity" resins or "specialty" resins. (The term resin dates from the early years of the plastics industry; it originally referred to naturally occurring amorphous solids such as shellac and rosin). Commodity resins are plastics that are produced at high volume and low cost for the most common disposable items and durable goods. They are represented chiefly by polyethylene, polypropylene, polyvinyl chloride, and polystyrene. Specialty resins are plastics whose properties are tailored to specific applications and that are produced at low volume and higher cost. Among this group are the so-called engineering plastics, or engineering resins, which are plastics that can compete with die-cast metals in plumbing, hardware, and automotive applications. Important engineering plastics, less familiar to consumers than the commodity plastics listed above, are polyacetal, polyamide (particularly those known by the trade name nylon), polytetrafluoroethylene (trademark Teflon), polycarbonate, polyphenylene sulfide, epoxy, and polyetheretherketone. Another member of the specialty resins is thermoplastic elastomers, polymers that have the elastic properties of rubber yet can be molded repeatedly upon heating.

Plastics also can be divided into two distinct categories on the basis of their chemical composition. One category is plastics that are made up of polymers having only aliphatic (linear) carbon atoms in their backbone chains. All the commodity plastics listed above fall into this category. The structure of polypropylene can serve as an example; here attached to every other carbon atom is a pendant methyl group (CH_3):

The other category of plastics is made up of heterochain polymers. These compounds contain atoms such as oxygen, nitrogen, or sulfur in their backbone chains, in addition to carbon. Most of the engineering plastics listed above are composed of heterochain polymers. An example would be polycarbonate, whose molecules contain two aromatic (benzene) rings:

The distinction between carbon-chain and heterochain polymers is reflected in the table, in which selected properties and applications of the most important carbon-chain and heterochain plastics are shown and from which links are provided directly to entries that describe these materials in greater detail. It is important to note that for each polymer type listed in the table there can be many subtypes, since any of a dozen industrial producers of any polymer can offer 20 or 30 different variations for use in specific applications. For this reason the properties indicated in the table must be taken as approximations.

Properties and applications of commercially important plastics					
Polymer Family and Type	Density (g/cm³)	Degree of crystal-linity	Glass transition tempera-ture (°c)	Crystal melting temperature (°c)	Deflection temperature at 1.8 mpa (°c)
Thermoplastics					
Carbon-chain					
High-density polyethylene (HDPE)	0.95–0.97	high	−120	137	—
Low-density polyethylene (LDPE)	0.92–0.93	moderate	−120	110	—
Polypropylene (PP)	0.90–0.91	high	−20	176	—
Polystyrene (PS)	1.0–1.1	nil	100	—	—
Acrylonitrile-butadiene-sty-rene (ABS)	1.0–1.1	nil	90–120	—	—
Polyvinyl chloride, unplasti-cized (PVC)	1.3–1.6	nil	85	—	—
Polymethyl methacrylate (PMMA)	1.2	nil	115	—	—
Polytetrafluoroethylene (PTFE)	2.1–2.2	moder-ate-high	126	327	—
Heterochain					
Polyethylene terephthalate (PET)	1.3–1.4	moderate	69	265	—
Polycarbonate (PC)	1.2	low	145	230	—
Polyacetal	1.4	moderate	−50	180	—
Polyetheretherketone (PEEK)	1.3	nil	185	—	—
Polyphenylene sulfide (PPS)	1.35	moderate	88	288	—
Cellulose diacetate	1.3	low	120	230	—
Polycaprolactam (nylon 6)	1.1–1.2	moderate	50	210–220	—
Thermosets*					
Heterochain					
Polyester (unsaturated)	1.3–2.3	nil	—	—	200
Epoxies	1.1–1.4	nil	—	—	110–250
Phenol formaldehyde	1.7–2.0	nil	—	—	175–300
Urea and melamine formaldehyde	1.5–2.0	nil	—	—	190–200
Polyurethane	1.05	low	—	—	90–100

Polymer family and type	Tensile strength (mpa)	Elonga-tion at break (%)	Flexural modulus (gpa)	Typical products and applications
Thermoplastics				
Carbon-chain				
High-density polyethylene (HDPE)	20–30	10–1,000	1–1.5	Milk bottles, wire and cable insulation, toys
Low-density polyethylene (LDPE)	8–30	100–650	0.25–0.35	Packaging film, grocery bags, agricultural mulch
Polypropylene (PP)	30–40	100–600	1.2–1.7	Bottles, food containers, toys
Polystyrene (PS)	35–50	1–2	2.6–3.4	Eating utensils, foamed food containers
Acrylonitrile-butadiene-sty-rene (ABS)	15–55	30–100	0.9–3.0	Appliance housings, helmets, pipe fittings
Polyvinyl chloride, unplasti-cized (PVC)	40–50	2–80	2.1–3.4	Pipe, conduit, home siding, window frames
Polymethyl methacrylate (PMMA)	50–75	2–10	2.2–3.2	Impact-resistant windows, skylights, canopies
Polytetrafluoroethylene (PTFE)	20–35	200–400	0.5	Self-lubricated bearings, nonstick cookware
Heterochain				
Polyethylene terephthalate (PET)	50–75	50–300	2.4–3.1	Transparent bottles, recording tape
Polycarbonate (PC)	65–75	110–120	2.3–2.4	Compact discs, safety glasses, sporting goods
Polyacetal	70	25–75	2.6–3.4	Bearings, gears, shower heads, zippers
Polyetheretherketone (PEEK)	70–105	30–150	3.9	Machine, automotive, and aerospace parts
Polyphenylene sulfide (PPS)	50–90	1–10	3.8–4.5	Machine parts, appliances, electrical equipment
Cellulose diacetate	15–65	6–70	1.5	Photographic film
Polycaprolactam (nylon 6)	40–170	30–300	1.0–2.8	Bearings, pulleys, gears
Thermosets				
Heterochain				
Polyester (unsaturated)	20–70	<3	7–14	Boat hulls, automobile panels
Epoxies	35–140	<4	14–30	Laminated circuit boards, flooring, aircraft parts
Phenol formaldehyde	50–125	<1	8–23	Electrical connectors, appliance handles
Urea and melamine formaldehyde	35–75	<1	7.5	Countertops, dinnerware
Polyurethane	70	3–6	4	Flexible and rigid foams for upholstery, insulation

Plastics are primarily defined not on the basis of their chemical composition but on the basis of their engineering behaviour. More specifically, they are defined as either thermoplastic resins or thermosetting resins.

Molding

Molding, also sometimes spelled moulding is the process of manufacturing by shaping liquid or pliable material using a rigid frame called a mold or matrix.

When molding plastics, a powder or liquid polymer such as polyethylene or polypropylene is placed into a hollow mold so the polymer can take its shape. Depending on the type of process used, various ranges of heat and pressure are used to create an end product.

Types of Plastic Molding

The most popular techniques in plastic molding are rotational molding, injection molding, blow molding, compression molding, extrusion molding, and thermoforming. We'll cover all these techniques in this guide to help you discover the best process to make your part or product.

Rotational Molding

Rotational Molding, also called rotomolding, is a manufacturing process for producing large hollow parts and products by placing a powder or liquid resin into a metal mold and rotating it in an oven until the resin coats the inside of the mold. The constant rotation of the mold creates centrifugal force forming even-walled products. Once the mold cools, the hardened plastic is removed from the mold.

Very little material is wasted during the process, and excess material is often re-used, making it economical and environmentally friendly.

Common uses for Rotational Molding

Rotational molding is commonly used to make large hollow plastic products like bulk containers, storage tanks, car parts, marine buoys, pet houses, recycling bins, road cones, kayak hulls, and playground slides.

Rotational Molds are Highly Customizable and Cost Effective

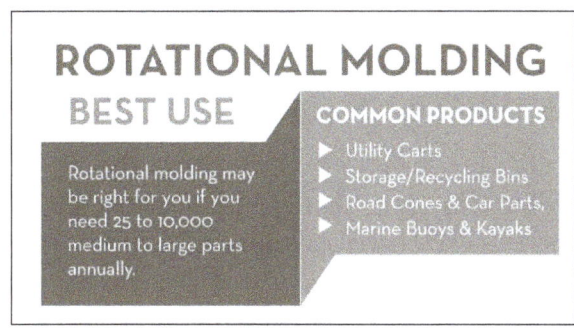

The mold itself can be highly intricate to facilitate the molding of a wide range of products. Molds can include inserts, curves, and contours as well as logos and slots for plastic or metal inserts to be placed after a product is molded.

Tooling costs are lower with rotational molds than injection or blow molds. The results are lower start-up costs and cost-effective production runs even when producing as few as 25 items at a time.

Injection Molding

Injection molding is the process of making custom plastic parts by injecting molten plastic material at high pressure into a metal mold. Just like other forms of plastic molding, after the molten plastic is injected into the mold, the mold is cooled and opened to reveal a solid plastic part.

The process is similar to a Jello mold which is filled then cooled to create the final product.

Common uses for Injection Molding

Injection molding is commonly used for making very high volume custom plastic parts. Large injection molding machines can mold car parts. Smaller machines can produce very precise plastic parts for surgical applications. In addition, there are many types of plastic resins and additives that can be used in the injection molding process, increasing its flexibility for designers and engineers.

Injection molds, which are usually made from steel or aluminum, carry a hefty cost. However, the cost per part is very economical if you need several thousand parts per year.

With injection molding, tooling usually takes 12-16 weeks with up to four more weeks for production.

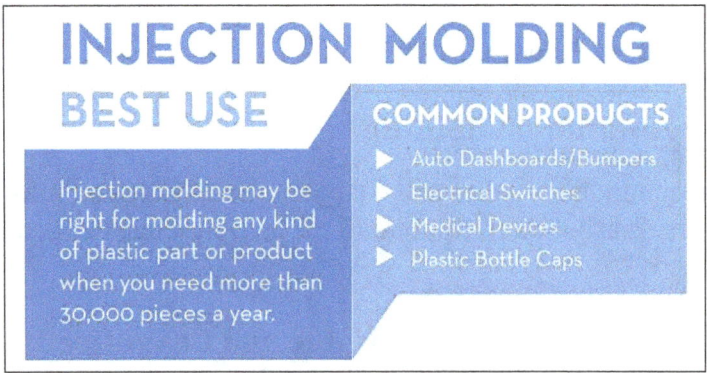

Blow Molding

Blow molding is a method of making hollow, thin-walled, custom plastic parts. It is primarily used for making products with a uniform wall thickness and where the shape is important. The process is based upon the same principle as glass blowing.

Blow molding machines heat up plastic and inject air blowing up the hot plastic like a balloon. The plastic is blown into a mold and as it expands, it presses against the walls of the mold taking its shape. After the plastic "balloon" fills the mold, it is cooled and hardened, and the part is ejected. The whole process takes less than two minutes so an average 12 hour day can produce around 1440 pieces.

Common uses for Blow Molding

Blow molding processes generate, in most cases, bottles, plastic drums, and fuel tanks. If you need a hundred thousand plastic bottles, this is the process for you. Blow molding is fast and economical with the mold itself costing less than an injection molding, but more than rotational molding sometimes as high as 6 to 7 times as much as a roto-molding tool.

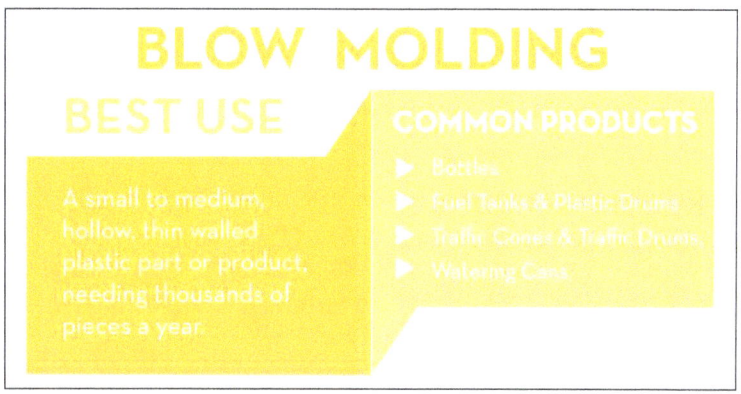

Compression Molding

Compression molding is done exactly like the name suggests. A heated plastic material is placed into a heated mold and then pressed into a specific shape. Usually, the plastic comes in sheets, but can also be in bulk. Once the plastic is compressed into the right shape, the heating process ensures that the plastic retains maximum strength. The final steps in this process involve cooling, trimming, and then removing the plastic part from the mold.

Common uses of Compression Molding

The best use of compression molding is the replacement of metal parts with plastic parts. It is mostly used for small parts and products in very high volume. The automotive industry uses compression molding heavily because the final products are very strong and durable.

The initial cost of a compression mold is substantial, depending on several factors including the number of cavities, the size of the parts, the complexity of the pieces, and the surface finish among other things. But the cost of each individual part is low at high quantities, so large quantities of parts are ideal for this form of molding.

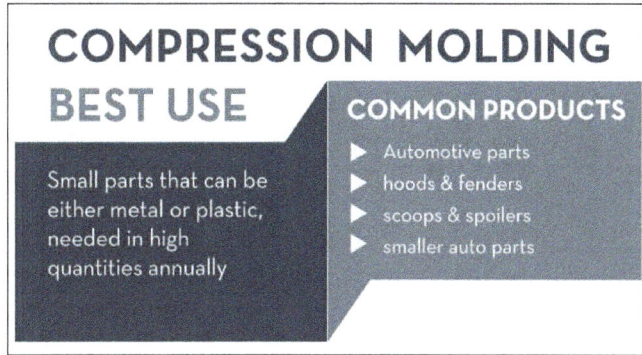

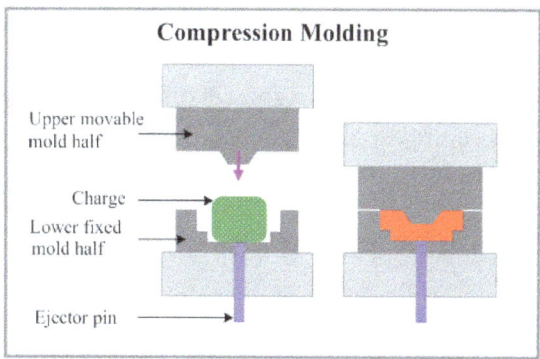

Extrusion Molding

Extrusion molding is similar to injection molding except that a long continuous shape is produced. Another difference in extrusion molding is that the process uses a "die" not a "mold."

Extruded parts are made by squeezing hot raw material through a custom die. A simplistic visualization would be like squeezing Play Doh through a shaped hole.

While other forms of molding use extrusion to get the plastic resins into a mold, this process extrudes the melted plastic directly into a die. The die shape, not a mold, determines the shape of the final product.

Common uses of Extrusion Molding

Parts made from extrusion have a fixed cross-sectional profile. Examples of extruded products include PVC piping, straws, and hoses. The parts do not need to be round but they need to have the same shape along the length of the part.

The cost of extrusion molding is relatively low compared to other molding processes because of the simplicity of the die and the machines themselves.

However, the nature of the extrusion molding process limits the kinds of products that can be manufactured with this technique.

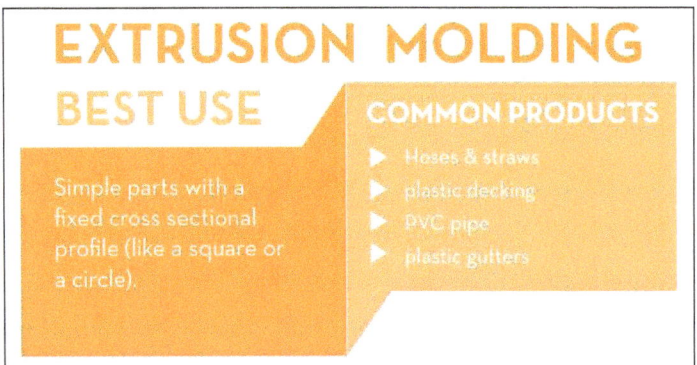

EXTRUSION MOLDING

BEST USE

Simple parts with a fixed cross sectional profile (like a square or a circle).

COMMON PRODUCTS
- Hoses & straws
- plastic decking
- PVC pipe
- plastic gutters

Thermoforming

Thermoforming is a manufacturing process where a plastic sheet called thermoplastic is heated to a pliable forming temperature, formed to a specific shape in a mold, and trimmed to create a usable product. Thermoplastic comes in a wide variety of materials, colors, finishes, and thickness.

Thermoforming uses several different types of molds and processes in order to achieve the final product. To create 3D products, the mold is typically a single 3D form made out of aluminum. Because thermoforming uses low pressures, molds can be produced for a low cost using inexpensive materials.

Common uses of Thermoforming

Thin-gauge thermoforming is commonly used to manufacture disposable cups, containers, lids, trays, blisters, clamshells, and other products for the food and general retail industries. Thick-gauge thermoforming includes parts as diverse as vehicle door and dash panels, refrigerator liners, and utility vehicle beds.

On average, it takes about 8 weeks to get a thermoform mold ready for production. The cost of a thermoform mold is based upon the size of the part that needs to be produced. A mold for a small part can cost as little as $20,000 while the cost of a larger mold can be upwards of $50,000.

Extrusion

Plastic extrusion is a process for converting plastic materials from solid to liquid states and reconstituting them as finished components. First, plastic pellets are gravity fed from a hopper into a jacketed screw. As the screw turns about its axis, it transports, melts, and pressurizes the plastic. From there, the molten material is forced through a die that shapes it into a specified cross-section, producing parts with a potentially wide range of lengths. During extrusion, plastics transform from solid to liquid and back again without sacrificing their distinctive properties. As a result, scrap parts can be ground and re-extruded with minimal degradation, making extrusion a popular method for reducing or recycling plastic waste.

Raw Thermoplastic Resins

Many plastic resin manufacturers sell both virgin and recycled goods made from extruded scrap that has been re-melted and returned to pellet form. These companies often purchase excess stock from production runs, obsolete parts, or unused resins for re-pelleting. This type of recycling can be a cost-effective and valuable method of eliminating industrial plastic waste.

In lieu of recycling, virgin thermoplastic resins can be purchased with laboratory certifications of purity, and standard technical grades are also available for general use. In addition, biodegradable plastics are increasingly prevalent, especially in blow-molded bottle production.

Resin manufacturers may add colorants, enhancers, or flock to their material in preparation for shipping. Subsequently, when fabrication companies receive the plastic stock, it is ready to be poured into the hoppers and extruded without little further pretreatment.

Single Screw Extrusion Machinery

There are numerous hardware considerations that can influence the quality of an extrusion. For example, screw geometry, screw rotation speed, and barrel heater temperature must be calibrated to suit the specific type of plastic being fabricated. Incompatible settings may hinder production or even damage the equipment.

As the main component of an extruder, the screw handles several tasks, including moving, melting, and pumping the plastic before it sends it through the die. A motor-driven gearbox with variable speeds usually turns the screw, which is enclosed in a tight fitting barrel. The mechanism is divided into three sections tailored to perform a sequence of specific tasks.

The feed section, located at the rear of the screw apparatus, contains a hopper that delivers resin pellets into the machine. As the screw turns, it draws the plastic forward with threads, or "flights." Barrel heaters help the plastic develop a tacky skin to improve friction between the plastic and the barrel wall. Without this friction, it would be difficult for the plastic to achieve lateral movement. As the plastic moves forward, it enters the transition, or melt, section. In this stage, the root diameter of the screw increases, while the flights decrease in size in order to melt the plastic by compressing and shearing it against the barrel wall.

The barrier screw is specifically designed with the transition section in mind. These screws have special barrier flights that improve mixing and melting by dividing the molten and solid plastic into separate channels. The barrier flights are smaller in diameter than regular flights, and provide a passage for melted plastic while blocking the solid pellets until they liquefy. As the plastic continues along the transition section, the melt channel increases in width, while the solid section decreases until there are no plastic pellets left.

After the plastic is melted and compressed, it is channeled into the metering section. Here, the plastic undergoes pressurized pumping, while the root diameter of the screw and the flight size remain constant. Some extrusion screws use special mixing heads to homogenize the plastic before it travels into the next section.

Extrusion Dies

The product assumes its final design inside the die. From the metering section, the plastic enters the front flange of the die, which is bolted onto the end of the extruder barrel. It flows around a metal parting tool, or "mandrel," suspended in the center of the channel. At the rear of the mandrel, a pin and a land size the product. The pin and the land are both removable, making it relatively simple to reconfigure the die or to replace worn parts.

Pressurized air is introduced into one of the mandrel's supports and exits from the die pin. This airflow prevents the product from collapsing as it leaves the die. Afterward, the component undergoes post-treatment.

Cooling and Sizing Equipment

When the product leaves the die, it enters a vacuum chamber where it is pulled through sizing rings. The combination of vacuum pulling and air pressure forces the plastic to conform to the shape of the rings. If the sizing rings become worn, they leave behind longitudinal scoring on the product. The vacuum chamber is water-filled, which cools the plastic into a hard solid. The cooled product is then pulled by belted runners and is cut to length or coiled up onto a spool.

Other Processes

Plastic is abundant in business and everyday life. You find it as product components, packaging and even in the containers used to ship products. For businesses interested in manufacturing or packaging their own products, an understanding of the major methods used in plastic manufacturing can prove useful in making decisions regarding which process best suits their purposes. Molding plastic to form specific products opens numerous avenues of invention for anyone willing to build the mold and create a functional and valuable product.

One of the most common plastic manufacturing methods, injection molding lends itself to mass production of products ranging from cell phone stands to toys. The injection molding process melts resin pellets inside the injection machine with a heated barrel. An auger moves the plastic forward and ensures an even mix of melted plastic. The machine then drives the melted plastic into a metal mold.

The plastic fills the mold and results in a solid plastic part or product. Most injection molding processes employ thermoplastics that you can melt and cool multiple times, which limits material waste.

Extrusion molding calls for a very similar process as injection molding. The machine still melts the plastic. Rather than filling a mold with the plastic, the machine presses the melted plastic through a die that gives the plastic a fixed shape. The extrusion molding process functions well in the production of a wide range of products, including pipes, door frames and seals.

The extrusion process can employ either multiple-melt thermoplastics or thermoset plastics, which only tolerate a single melting cycle.

Manufacturing with Blow Molding

Several variations of the blow molding process exist. The essential process calls for the production of a hollow, pre-shaped length of melted thermoplastic, known as a parison. A mold closes around the parison. Air pressure forces the hollow plastic to expand into the mold shape, leaving the interior of the object hollow.

Variations on the blow molding process include injection and extrusion blow molding as well as stretch blow molding. Manufacturers employ blow molding processes to make bottles and other containers.

Manufacturing with Rotational Molding

Rotational molding offers a second option for manufacturing hollow objects. In rotational molding, the plastic powder goes in the mold before heating. The closed mold enters a furnace and rotates, which allows the plastic powder to coat the entire interior of the mold. The heat melts the plastic into a single layer that conforms to the shape of the mold cavity, while leaving the interior of the final product hollow.

Manufacturers use rotational molding to create products and components for a wide range of uses, including auto parts, toys and furniture.

Compounding

The first step in most plastic fabrication procedures is compounding, the mixing together of various raw materials in proportions according to a specific recipe. Most often the plastic resins are supplied to the fabricator as cylindrical pellets (several millimetres in diameter and length) or as flakes and powders. Other forms include viscous liquids, solutions, and suspensions.

Mixing liquids with other ingredients may be done in conventional stirred tanks, but certain operations demand special machinery. Dry blending refers to the mixing of dry ingredients prior to further use, as in mixtures of pigments, stabilizers, or reinforcements. However, polyvinyl chloride (PVC) as a porous powder can be combined with a liquid plasticizer in an agitated trough called a ribbon blender or in a tumbling container. This process also is called dry blending, because the liquid penetrates the pores of the resin, and the final mixture, containing as much as 50 percent plasticizer, is still a free-flowing powder that appears to be dry.

The Banbury mixer used for the mixing of polymers and additives in the manufacture of plastic and rubber.

The workhorse mixer of the plastics and rubber industries is the internal mixer, in which heat and pressure are applied simultaneously. The Banbury mixer resembles a robust dough mixer in that two interrupted spiral rotors move in opposite directions at 30 to 40 rotations per minute. The shearing action is intense, and the power input can be as high as 1,200 kilowatts for a 250-kg (550-pound) batch of molten resin with finely divided pigment.

In some cases, mixing may be integrated with the extrusion or molding step, as in twin-screw extruders.

Fiber

Fibers are filiform elements, which present a high length ratio with respect to their maximum transverse dimension. They are characterized by flexibility and fineness.

Fibers are made of macromolecules referred as polymers. In turn, these polymers are composed of a sequence of monomers. Polymers are chemically stable while monomers are chemically unstable, which explains the reaction of the union of monomers in the formation of a polymer.

The polymer length is a very important factor, since almost all fibers have very long polymer chains. Regarding the molecular arrangement, fibers can be highly or slightly oriented. When they are highly oriented, they conform a crystalline region, which means that the polymers are longitudinally aligned and in order, more or less parallel. In the case of fibers being slightly oriented, amorphous regions are formed, where the polymers do not have a defined orientation.

Highly orientated polymers confer fibers high tensile strength, low elongation, heat resistance and chemical resistance. On the contrary, the amorphous areas of slightly oriented polymers give fibers features such as: flexibility, softness and comfort.

Fibers can be classified attending to several aspects:

- With respect to the length, fibers can be classified as discontinuous, when they are limited to a few centimeters length; or continuous, when they have a very high length, which is only limited by technical reasons;

- With respect to its origin, they can be classified as natural or man-made fibers. Within the latter group, artificial, synthetic and inorganic fibers can be found. Natural fibers exist as they are found in nature, and can be of animal, vegetable or mineral nature.

Textile and Fabrics Properties

A unit of matter, either natural or manufactured, that forms the basic element of fabrics and other textile structures. A fiber is characterized by having a length at least 100 times its diameter or width. The term refers to units that can be spun into a yarn or made into a fabric by various methods including weaving, knitting, braiding, felting, and twisting. The essential requirements for fibers to be spun into yarn include a length of at least 5 millimeters, flexibility, cohesiveness, and sufficient strength. Other important properties include elasticity, fineness, uniformity, durability, and luster.

Properties of Fiber

We saw three types of properties according to the fiber properties for a textile fiber. There are a lot of characteristic in the textile fibers. But it is characterized as three basic characterization.

They are:

- Physical properties,

- Chemical properties,

- Mechanical properties of Textile fiber.

1. Physical Properties:

- Length ----------Staple (15mm - 150 mm).

- Fineness ---------- Length: Width = 1000:1.

- Cross Sectional Shape.

- Crimp.

- Density.

2. Mechanical Properties:

- Strength (Tenacity) (P.S.I).

- Elasticity (Recovery percentage).

- Extensibility (Breaking Extension).

- Rigidity (Stiffness).

3. Chemical Properties: Solubility in aquas and organic solvent. Useful properties of another hind desired in a textile fiber are indicated below.

- Behavior towards dyes.

- Ability to moisture absorption.

- Resistance to deteriorating influence including; light, thermal stability, resistance to bacteria, mildow moth and other destructive insect, corrosive chemicals.

Spinning

Fibers are formed by the extrusion of the polymer melt or spin dope through tiny holes in a spinneret plate. Such a plate may contain 1,000 holes or more. Textile fibers are relatively fine, so the diameter of the hole may be only a few mils, where one mil is 0.001 inches (25.4 μm). The thickness of the filament is generally not given in linear dimensions, but rather in terms of mass per length. For some reason the fiber industry has adopted the terms denier and denier per filament, dpf to express the filament size. One dpf corresponds to a mass of 1 g in a length of 9000 meters.

If the density of the polymer is 1 g/cm3, this would correspond to a diameter of 1.2 x 10⁻³ cm, or about half a mil. Typically, textile fibers are in the range of 3 to 15 dpf. Recall that one g is roughly 1/30 of an ounce.

In melt spinning, the filaments are normally drawn down, or stretched, just downstream of the spinneret holes. The stretch is of the order of 2 to 3x, so the spinneret hole may be 50 to 75% larger than the filament diameter when it is first cooled. Additional post formation stretching may also be used, however, so that the final filament diameter may be one-half or less than the diameter of the spinneret hole.

The spinneret hole is usually only slightly longer than its width, in part to minimize pressure drop at the plate. But the plate still has to be strong enough to withstand the upstream pressure. For this reason, the melt passes through a conical section before reaching the final spinneret hole, so that the plate can be relative thick. Pressure drop through this converging section is very difficult to calculate for these polymeric materials, because the extensional flow rheology is usually not well characterized. One can readily visualize the alignment of the polymer molecules in the converging section, where the polymer undergoes a severe stretching step. Ignoring this pressure loss, let us focus on the spinneret hole itself.

An important question here is whether one can use the Hagen-Poiseuille equation to compute the pressure drop in such a short tube. To help answer this question, we first compute the entrance length, which is approximately equal to the axial distance downstream from a tube entrance at which the momentum boundary layers merge at the center axis, where a fully parabolic profile is established. This distance, from Bird, Stewart, and Lightfoot, Transport Phenomena, is:

$$L_e/D = 0.035 \, Re.$$

Where L_e is the entrance length, D is the tube diameter and Re is the Reynolds number.

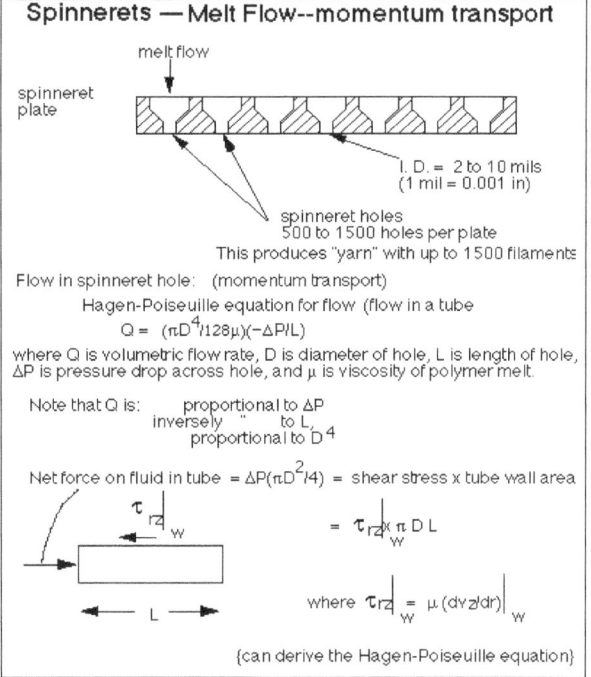

Flow in spinnerets

For a representative calculation, we consider nylon melt, with a viscosity of 200 Poises, being spun from a hole 10 mils in diameter at a final spinning speed of 2,000 ypm and a stretch of 5x between the spinneret and the final take-up. This means that the bulk velocity, ub, in the spinneret hole is 400 ypm. The Reynolds number, for a specific gravity of about one, is:

Re = ub D/n = 0.077

So, the entrance length is less than 3 thousandths of the diameter of the hole. Therefore, we can safely use the Hagen-Poiseuille equation to calculate the pressure drop. The equation is:

$$\frac{\Delta P}{L} = -\frac{32\mu u_b}{D^2}$$

For the nylon example we just explored, the pressure drop is predicted, for a length of 3.0 mils, to be 2200 psi. This pressure drop might be a bit excessive in practice, but the method of calculation remains illustrative. One part of the calculation which was not taken into account is the power-law behavior of most polymer melts and polymer solutions. Such behavior usually is revealed by a shear-thinning response, in which the apparent viscosity decreases as shear rates increase. This would lead to significantly lower pressure drops for the spinneret plate.

As the polymer exits the spinneret hole, it tends to swell and this swell is especially noticeable at low filament tensions. Apparently, the polymer molecules must coil under the shearing action within the hole and, as it exits, the polymer molecules are free to uncoil, as seen by an expansion of the polymer stream jetting out of the hole. This phenomenon is referred to as "die swell," and can even amount to a doubling or more of stream diameter. Even Newtonian fluids can be shown to exhibit a swelling at the exits of tubes, even at very low Re; the predicted extent of swell is about 14% for Newtonian fluids. Since, in fiber spinning, the filaments are under tension, the extent of die swell is considerably reduced. Furthermore, the extent of swell appears to have no influence of final filament properties. In order for the filaments to undergo stretching, some power must go into the stretching motion immediately downstream of the spinneret plate, but the amount of this power is negligible.

Melt Spinning

In the spinning of molten polymers, such as nylon, polyester, and polypropylene, melt spinning begins with a cooling of the molten filament after it leaves the spinneret. At the same time, the filament is pulled downwards towards the take-up section and this resulting tension in the molten filament provides a stretching action in the molten filament itself. In most melt spinning operations the degree of stretch is of the order of 3x, which means that the velocity of the initially cooled, or solid, fiber is about three times the average velocity of the melt coming out of the spinneret. For some filaments, this initial stretch is very important in helping to establish properties in the polymer which depend on whether one deals with the properties in the fiber axis direction or in the fiber radius direction. This directional dependence of properties is called anisotropy and the usual example is that of a slab of wood, in which strength and fracture properties along the grain are quite different from properties across the grain. (With many fibers, however, these properties are controlled downstream, where the fibers are reheated, stretched further, and cooled again).

In any case, the polymer melt, once it comes out of the spinneret hole, starts to cool down and also starts to stretch out. Because the "apparent viscosity" of the melt increases rapidly as the melt cools, most of the stretching takes place in a region relatively close to the spinneret hole whereas "most" of the cooling takes place well away from this hole. But these terms and descriptions are not exact and are not easily quantified. The real advantage in using these descriptions is that it permits us to make a simplifying assumption as we analyze the melt spinning process. The assumption is: We can separate the stretching and cooling operations into two separate distinct regions, with the first occurring relative close to the spinneret (say, within 10% of the distance to the first take-up, or speed-control roll), and the second over the remaining distance to the first take-up roll. If need be, we could return later to this assumption to determine its degree of accuracy, but let us accept it for the moment.

The stretching region, within which the relatively long polymer molecules become aligned along the filament axis, might be characterized by very complex rheology. Within the field of polymer processing, rheology deals with the relationship between stress and the history of strain; for Newtonian fluids, you can recall that the fluid stress is proportional to the rate of strain (the shear rate). We are not really too concerned about the polymer melt rheology here, however, since it will not likely be important in determining the power required for the first drive or take-up roll. Frictional and interfacial stresses are likely to be far more important. Therefore, in terms of design considerations, we can probably ignore that part of the melt-spinning process in which the initial, post-spinneret stretching of the polymer melt occurs and focus instead on the cooling step of the melt spinning operation.

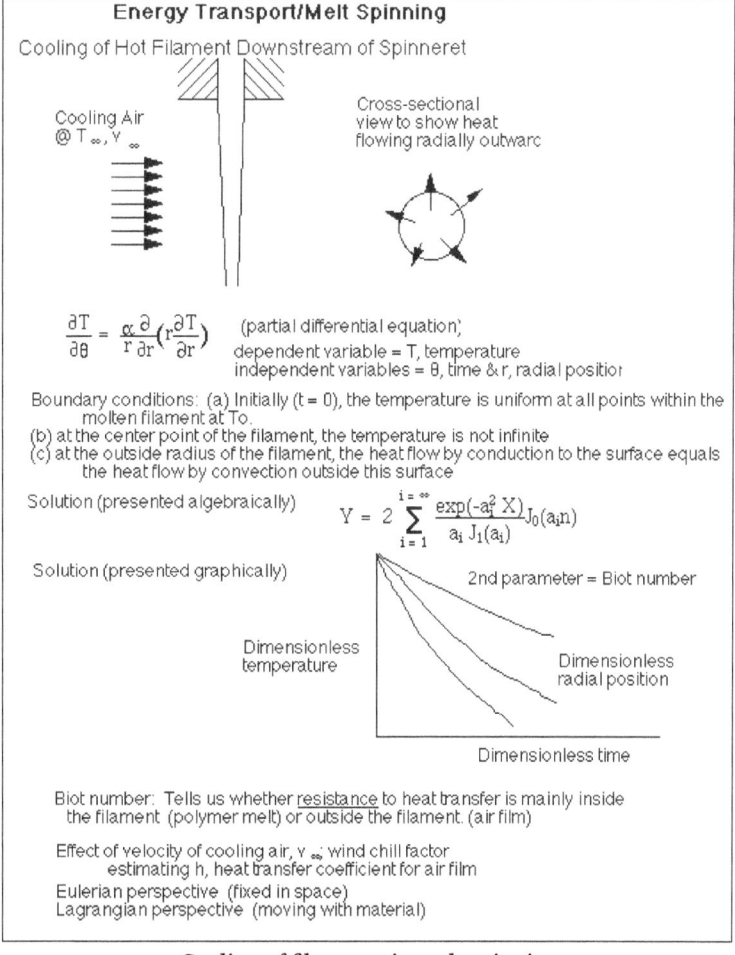

Cooling of filaments in melt spinning

Perhaps the most important design consideration in the melt spinning process is the cooling of the filaments. In order to simplify our analysis, we restrict our focus to the point where the filaments have reached a uniform diameter (recall that we previously asserted that this is a relatively short distance) and that they are at some initial temperature which will be somewhat cooler (by about 20 °C) than the melt temperature at the spinneret exit. At this point, the temperature within the filament will depend on radial position, with the maximum occurring at the center, on the filament axis. We shall invoke an approximation of a flat temperature profile, in which the temperature does not vary with r, in order to utilize existing mathematical solutions. An important part of learning engineering is to learn how to take "appropriate shortcuts" which save time with little sacrifice in accuracy. This is one example. The melt spinning process is steady: viewing the spinning thread-lines at a fixed position (the socalled Eulerian perspective) shows that nothing appears to change with time. If one situates oneself on the moving threadline (figuratively, of course) there does certainly appear to be a time dependence to the temperature of the filament. This viewpoint of moving with the material is called the Lagrangian perspective. Whereas the Eulerian perspective requires you to measure the threadline temperature as a function of r, radial position within the filament, and z, axial position along the filament, in order to follow the cooling, the Lagrangian perspective allows you to follow the cooling as a function of r and t, where t is time. Zero time should be some convenient reference-here it would correspond to locating yourself on the filament at the end of the stretching region and at the "beginning" of the cooling region. This is equivalent to the cooling of an infinite rod which is fixed in space. The governing differential equation, which can be derived easily using shell balance techniques is:

$$\frac{\partial T}{\partial \grave{e}} = \frac{\alpha \partial}{r \partial r}\left(r\frac{\partial T}{\partial r}\right)$$

Where q is time and a is the thermal diffusivity of the polymer. The student will readily recognize this as a partial differential equation, since the temperature T depends on both r and q. In order to solve the equation quantitatively, one must specify initial and boundary conditions. The boundary is naturally R the outside radius of the filament. So the initial condition (q = 0) is simply:

$$T(r,q) = T_0 \text{ for } r < R$$

Where T_0 is a constant. We need two boundary conditions, corresponding to r = 0 and r = R. At r = 0, T remains finite; although some prefer to say:

$$\frac{\partial T}{\partial r} = 0$$

based on symmetry arguments. At r = R, the heat arriving at the surface by conduction from within must match the heat leaving by convection:

$$-k\frac{\partial T}{\partial r} = h(T - T_{oo})$$

at r = R and all q > 0. k is the thermal conductivity of the filament (we shall assume that this conductivity does not change as the polymer solidifies. h is the heat transfer coefficient governing the heat transfer from the surface to the surrounding air. h can be estimated from various correlations if information about the velocity and direction of the cooling air is given. An example of such a correlation for heat transfer from a cylinder in crossflow is given by Churchill and Bernstein:

$$Nu_D = 0.3 + \frac{0.62\,Re_D^{1/2}\,Pr^{1/3}}{[1+(0.4/Pr)^{2/3}]^{1/4}}[1+(\frac{Re_D}{28\,200})^{5/8}]^{4/5}$$

Where NuD is the Nusselt number, hD/k, (k is the thermal conductivity of the fluid in crossflow and D is the cylinder diameter) and ReD is the Reynolds number based on the fluid in crossflow. Pr is the Prandtl number, n/a and is also based on the crossflow fluid. The student will recall that an important advantage of presenting correlations in terms of dimensionless variables like Nu (dimensionless heat transfer coefficient) and Re (inertial stresses divided by viscous stresses) is that the resulting expression is often simpler, revealing more clearly the relationships among such variables.

We shall also assume that the heat of fusion is negligible, primarily because we want to simplify the calculation. This assumption, especially for crystalline polymers, could be very poor, however, and could lead an underestimate of the cooling time by a factor of two.

As engineers, or even as normal, sane people, we would not want to solve the differential equation for every single set of geometries, thermal diffusivities and initial and boundary conditions. We can avoid this needless energy expenditure if we express the differential equation in dimensionless form:

$$\frac{\partial Y}{\partial X} = \frac{\partial^2 Y}{\partial n^2}$$

where

$$Y = \frac{T_{oo}-T}{T_{oo}-T_o}\quad n = \frac{r}{R}, and\, X = \frac{\alpha\theta}{R^2}$$

Y is called the unaccomplished temperature change, since it starts at unity at time zero and declines from there. n is the normalized radial position, and X is dimensionless time, sometimes called the Fourier number. One final dimensionless group, m, expresses the relative resistance outside the filament to that in the filament:

$$m = \frac{k}{h\,R}$$

Finally, the initial and boundary conditions become:

Y = 1 at X = 0 and 0 < n < 1;

$$\frac{\partial Y}{\partial n} = 0\, at\, n = 0\, and\, X > 0.$$

$$-m\frac{\partial Y}{\partial n} = Y\, at\, n = 1\, and\, X > 0.$$

The solution, Y(n, X) is then valid for any case of unsteady-state heat conduction within a cylindrical geometry with a uniform initial temperature and convective heat transfer from the surface to a surrounding fluid at a uniform temperature T . The solution is shown in graphical form on the slide and is available in almost all transport textbooks. The resulting charts are known variously as "Gurney-Lurie Charts" or "Heissler Charts," depending on which reference or form of charts you use. An analytical solution for a slightly less general case is given below. Note that, by use of dimensionless variables, we have successfully created a result which is applicable to a broad range of geometries and material properties. For the special case in which heat transfer resistance from the surface of the fluid to the surrounding fluid is negligible, one can set h, the heat transfer coefficient, to (this, of course, is equivalent to setting the temperature of that surface to that of the surrounding fluid for all q > 0) and the analytical solution is:

$$Y = 2\sum_{i=1}^{i=\infty} \frac{\exp(-a_i^2 X)}{a_i J_1(a_i)} J_0(a_i n)$$

Where J_0 and J_1 are Bessel functions of the first kind (zero and first order, respectively), and ai is the ith root of $J_0 (ai) = 0$. This solution is presented in the transport text Momentum, Heat, and Mass Transfer, by Bennett and Myers.

Dry Spinning

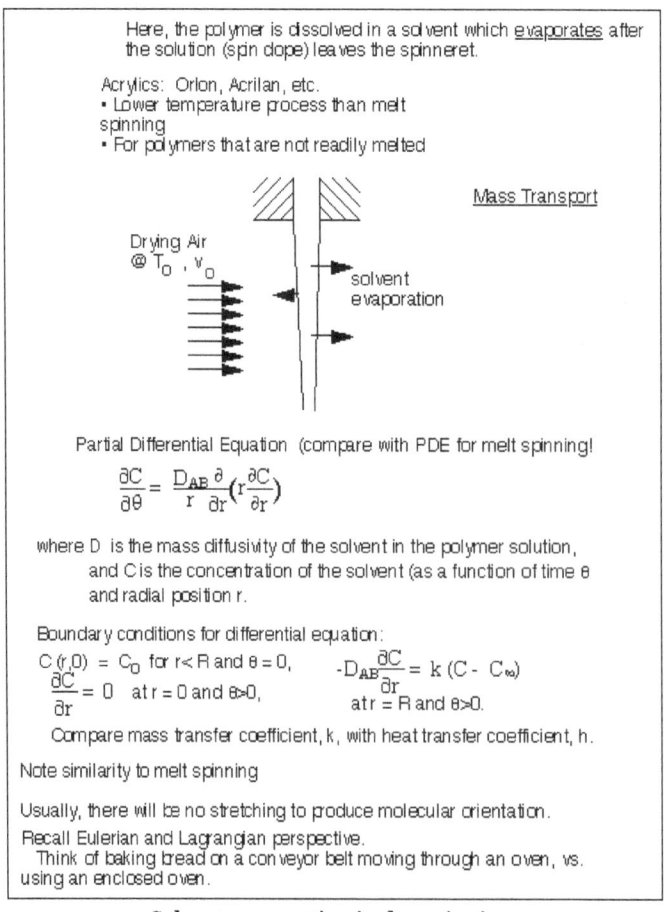

Solvent evaporation in dry spinning.

Unlike melt spinning, both dry and wet spinning use solvents in which the polymer dissolves. The resulting solution or suspension is a viscous "spin dope." This process necessarily introduces another species, which is subsequently removed, and therefore is more expensive than conventional melt spinning processes. It is used in cases where the polymer may degrade thermally if attempts to melt it are used or in cases where certain surface characteristics of the filaments are desired-melt spinning produces filaments with smooth surfaces and dry spinning produces filaments with rough surfaces. The rougher surface may be desirable for improved dyeing steps or for special yarn characteristics.

The term "dry spinning" is a bit misleading, since the polymer is certainly wet by a solvent. Presumably, the intent here was to distinguish the two methods of solvent removal for the two cases of dry and wet spinning. The solvent in dry spinning is a volatile organic species and this solvent starts to evaporate after the filament is formed, which is immediately downstream of the spinneret. Whereas melt spinning involved solidification by cooling, dry spinning produces solidification of the polymer by solvent removal.

Several commercial fibers, including acrylic fibers such as Orlon, are made by a dry spinning process. You may recall that these acrylic fibers are popular as substitutes for wool fibers. In any case, the spinning step which defines, in large part, the spinning process is that of solvent removal from the filaments. In the case of Orlon, the polymer, polyacrylonitrile, is dissolved to a polymer concentration of 20 to 30 wt% in a dimethylformamide solvent. Warm gases (air? - probably not, on account of the need for solvent recovery) are passed through the fiber bundle in the region just downstream of the spinneret. This begins to look very much like the cooling crossflow in melt spinning. The solvent encounters both a diffusional resistance within the fiber and a convective resistance in moving from the surface of the filament to the crossflow gases. Within the filament, the material property of greatest importance is DAB, the diffusivity of the solvent A through the filament B. Here, we can characterize the diffusive flux of the solvent by:

$$N_A = -D_{AB} \frac{dC_A}{dr}$$

which is your familiar Fickian Diffusion equation. We use the ordinary derivative here because the process is steady and we have not yet begun to use the Lagrangian perspective. One point to emphasize here is the similarity of this equation to the Fourier heat conduction equation. If we then adopt the Lagrangian perspective, we have:

$$\frac{\partial C}{\partial è} = \frac{D_{AB}}{r} \frac{\partial}{\partial r} (r \frac{\partial C}{\partial r})$$

Comparison with the unsteady heat conduction equation reveals the equation to be identical with the exception that a is replaced by DAB and T by C. Both a and DAB have dimensions of length squared over time or units of cm^2/s. The initial and boundary conditions are also practically identical to the heat transfer case, with the assumption of uniform concentration profile at time zero, and zero concentration gradient on the filament centerline and matching diffusive and convective flux at the filament surface:

C (r,0) = Co for r< R and q = 0,

$$\frac{\partial C}{\partial r} = 0 \text{ at } r = 0 \text{ and } q > 0$$

and

$$-D_{AB}\frac{\partial C}{\partial r} = k(C - C_{oo}) \text{ at } r = R \text{ and } q > 0.$$

Instead of heat transfer coefficient, h, we have mass transfer coefficient, k. Correlations for k, expressed in terms of a dimensionless mass transfer coefficient, Sh (for Sherwood number) as a function of ReD and Sc, are also available. Sc is the ratio of momentum diffusivity to mass diffusivity, n/DAB, (for the cross flow fluid) and is comparable to the Prandtl number, n/a. Note that n and DAB are the momentum diffusivity and mass diffusivity of the gas in crossflow, and not of the polymer solution. One correlation for k is:

Sh = 0.281 (ReD)-0.4

Where Sh = (kD/DAB). This correlation is a bit unusual, in that normally the mass transfer coefficient is found to be proportional to the diffusivity raised to a power less than unity. The reader may want to check other correla tions to check whether this form may or may not be reasonable.

Just as we dedimensionalized the heat transfer equations, we can do the same for solvent diffusion. The resulting equations then are exactly identical to those for unsteady heat conduction.

$$\frac{\partial Y}{\partial X} = \frac{\partial^2 Y}{\partial n^2}$$

$$Y = \frac{C_{oo} - C}{C_{oo} - C_o}, n = \frac{r}{R}, \text{ and } X = \frac{D_{AB}\,\theta}{R^2}$$

$$m = \frac{D_{AB}}{kR}.$$

Of course, T is replaced by C, a by D_{AB}, h by k. Therefore, the same solution (graphical or analytical) is obtained and the same charts can be used to obtain quantitative predictions of the fiber spinning process. One can readily calculate, therefore, the Fourier number, X, required for the solvent concentration at the filament centerline to become less that 1% of the original value (Y < 0.01). From this value for X, the actual time (in a Lagrangian sense, remember) can be calculated. Finally, by multiplying this time by the yarn speed, the length of the solvent recovery section is obtained directly. The analogy here might be that of using a conveyor belt in a tunnel oven to bake bread. We can calculate the length of the tunnel oven, once we know the time to bake the bread and the speed of the conveyor belt.

Wet Spinning

Fibers produced by wet spinning include rayon and Kevlar™. Rayon was originally developed as a synthetic substitute for silk and Kevlar was produced as a high-strength fiber for use in various aerospace and specialty-use applications. Furthermore, many comercial acrylic fibers are also produced by wet spinning.

As with dry spinning, the polymer is dissolved or suspended in a solvent, to form a viscous "spin dope" and filaments are formed by extrusion through tiny holes in a spinneret plate. Kevlar, for example, will degrade thermally if attempts are made to melt it, and thus a solvent must be used. With wet spinning, the term more accurately depicts the process than with dry spinning, because the solvent is extracted or, perhaps more appropriately, leached, from the filaments by another liquid. In most cases, the second liquid is aqueous.

A major difference between wet spinning and either melt or dry spinning is that one is spinning into a fluid (liquid) with a much higher viscosity. Because this higher viscosity can translate into high shearing stresses on the surfaces of the filaments, the tension in the filaments can become quite high. For example, towing a buoy by a long line behind a boat can produce very high tensions in the line when compared with towing the same buoy by a short line. For long baths, the tension can become sufficiently high that the filaments might break, as their tensile strength is exceeded. To avoid this danger, much lower spinning speeds must be used. Whereas melt spinning may utilize spinning speeds of 2,000 yards per minute (80 mph), spinning speeds in wet spinning are usually less than 300 ypm.

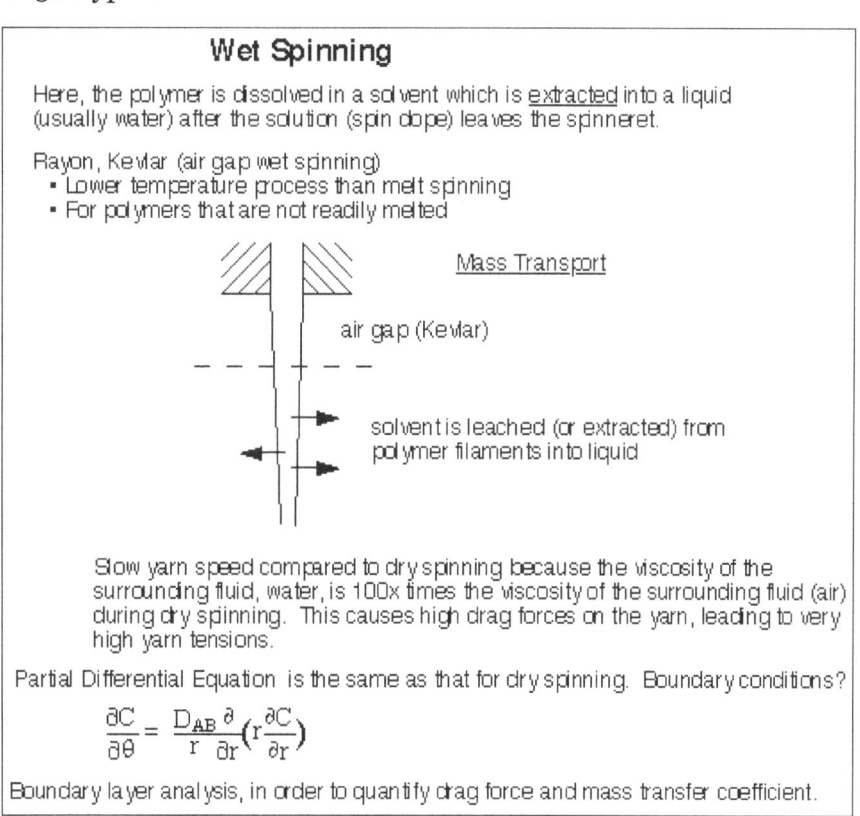

Wet Spinning- Solvent Removal by Extraction or Leaching.

Another difference with dry spinning is the capability of using many more spinneret holes in the case of wet spinning. The total number can approach 60,000 in a single spinneret plate, if the spinning is done directly into a coagulating or extracting liquid. Because the liquid is present, the filament forms a type of skin almost immediately and the potential for the filaments to touch and fuse is practically eliminated, compared with dry or melt spinning.

In the case of Kevlar the spin dope is relatively warm, about 100°C, and forms a viscous, liquid crystal. The solvent is sulfuric acid, at a concentration of about 80 wt% (20 wt% polymer). These liquid crystals are easily oriented by a stretching motion. Therefore, during the spinning process, the filaments are first extruded through an air gap, where the filaments undergo strains of 2 to 3x, which produces a high degree of molecular orientation in the filaments. This air gap is of the order of one inch. It also allows the spinneret plate to be warm (100 °C) while the extraction bath can be cool (ca 15 °C). The hot filaments then strike the cooling bath where the filaments are "quenched" and much of the orientation is locked in by the rapid cooling action. Subsequent to the quench step, the solvent is extracted, which requires a relatively long bath contact time. But the initial quenching step is crucial, since it allows for the oriented molecules to be "frozen" into position. This orientation is particularly important to the high-strength properties of Kevlar-the filaments, on a weight basis, are stronger that steel, almost by a factor of two. If one attempts to use the same process to produce Kevlar filaments of large diameter, the core of the filaments can lose its orientation, because the quench time to reach the core will increase with the square of the filament radius. The filament skin, or the outer part of the filaments, however, will have the orientation locked in and a high degree of orientation will exist there. This produces a so-called "skin-core" effect, in which the average properties of the filaments, expressed as tensile strength per unit cross-sectional area, will decline on account of a decreased average orientation.

Kevlar, with its focus on strength development via "air-gap" wet spinning, is somewhat unique within the process of wet spinning. As with melt- and dry- spinning, the controlling part of the process is associated with development of the filament structure, either by cooling of the filament or by removal of the solvent. The equations for diffusion in wet spinning are identical to those for dry spinning, with the exception that the fluid passing outside the filaments is a liquid and not a gas. Also, the flow may not be across the filaments, but even, partially, along the filaments. Therefore, the correlations and nature of the flow surrounding the filaments will result in different values for the surface mass transfer coefficient. Whether this will change the relative resistance dramatically will depend on the particular fiber to be produced and its dimensions and properties. The same graphical solutions described earlier can be used, however.

To design a wet-spinning process, it may be necessary to predict the transport of momentum, heat, and mass in the region adjacent to the filament just downstream of the spinneret. One can use a so-called "boundary-layer" analysis to do this. Treatment of such an analysis is beyond the scope of the present discussion of fiber spinning, but a brief description of the analysis is appropriate. One form of boundary-layer analysis involves von Karman integral boundary-layer techniques. The boundary layer starts at zero thickness at the first point where the fiber contacts the extracting liquid and growing gradually radially outwards from each filament as one proceeds downstream. The velocity profile inside the boundary layer is assumed and all of the velocity change between the filament and the surrounding fluid is contained within this "momentum" boundary layer. Similarly, thermal and diffusional boundary layers contain all the changes in temperature and concentration, respectively. Based upon approximations of these velocity profiles, frequently assumed to be turbulent, the variation in filament drag with position can be predicted, along with local heat and mass transfer coefficients. The student is referred to Transport Phenomena, by Bird, Stewart, and Lightfoot, for additional details of such integral boundary-layer technique.

Elastomer

Elastomer is any rubbery material composed of long chainlike molecules, or polymers, that are capable of recovering their original shape after being stretched to great extents—hence the name elastomer, from "elastic polymer." Under normal conditions the long molecules making up an elastomeric material are irregularly coiled. With the application of force, however, the molecules straighten out in the direction in which they are being pulled. Upon release, the molecules spontaneously return to their normal compact, random arrangement.

The elastomer with the longest history of use is polyisoprene, the polymer constituent of natural rubber, which is made from the milky latex of various trees, most usually the *Hevea* rubber tree. Natural rubber is still an important industrial polymer, but it now competes with a number of synthetics, such as styrene-butadiene rubber and butadiene rubber, which are derived from by-products of petroleum and natural gas.

Polymers and Elasticity

A polymeric molecule consists of several thousand chemical repeating units, or monomers, linked together by covalent bonds. The assemblage of linked units is often referred to as the "chain," and the atoms between which the chemical bonding takes place are said to make up the "backbone" of the chain. In most cases polymers are made up of carbon backbones—that is, chains of carbon (C) atoms linked together by single (C$-$C) or double (C=C) bonds. In theory, carbon chains are highly flexible, because rotation around carbon-carbon single bonds allows the molecules to take up many different configurations. In practice, however, many polymers are rather stiff and inflexible. The molecules of polystyrene (PS) and polymethyl methacrylate (PMMA), for instance, are made up of relatively bulky units so that, at room temperature, free motion is hindered by severe crowding. In fact, the molecules of PS and PMMA do not move at all at room temperature: they are said to be in a glassy state, in which the random, "amorphous" arrangement of their molecules is frozen in place. All polymers are glassy below a characteristic glass transition temperature (T_g), which ranges from as low as -125 °C (-195 °F) for an extremely flexible molecule such as polydimethyl siloxane (silicone rubber) to extremely high temperatures for stiff, bulky molecules. For both PS and PMMA, T_g is approximately 100 °C (212 °F).

Some other polymers have molecules that fit together so well that they tend to pack together in an ordered crystalline arrangement. In high-density polyethylene, for example, the long sequences of ethylene units that make up the polymer spontaneously crystallize at temperatures below about 130 °C (265 °F), so that, at normal temperatures, polyethylene is a partially crystalline plastic solid. Polypropylene is another "semicrystalline" material: its crystallites, or crystallized regions, do not melt until they are heated to about 175 °C (350 °F).

Thus, not all polymers have the necessary internal flexibility to be extensible and highly elastic. In order to have these properties, polymers must have little internal hindrance to the random motion of their monomer subunits (in other words, they must not be glassy), and they must not spontaneously crystallize (at least at normal temperatures). On release from being extended, they must be able to return spontaneously to a disordered state by random motions of their repeating units as a result of rotations around the carbon-carbon bond. Polymers that can do so are called elastomers.

Four common elastomers are *cis*-polyisoprene (natural rubber, NR), *cis*-polybutadiene (butadiene rubber, BR), styrene-butadiene rubber (SBR), and ethylene-propylene monomer (EPM). SBR is a mixed polymer, or copolymer, consisting of two different monomer units, styrene and butadiene, arranged randomly along the molecular chain. EPM also consists of a random arrangement of two monomers—in this case, ethylene and propylene. In SBR and EPM, close packing and crystallinity of the monomer units are prevented by their irregular arrangement along each molecule. In the regular polymers NR and BR, crystallinity is prevented by rather low crystal melting temperatures of about 25 and 5 °C (approximately 75 and 40 °F), respectively. In addition, the glass transition temperatures of all these polymers are quite low, well below room temperature, so that all of them are soft, highly flexible, and elastic. The principal commercial elastomers are listed in the table, which also indicates some of their important properties and applications.

Properties and Applications of Commercially Important Elastomers						
Polymer type	Glass transition temperature (°C)	Melting temperature (°C)	Heat resistance	Oil resistance	Flex resistance	Typical products and applications
Polyisoprene (natural rubber, isoprene rubber)	−70	25	P	P	E	Tires, springs, shoes, adhesives
Styrene-butadiene copolymer (styrene-butadiene rubber)	−60		P	P	G	Tire treads, adhesives, belts
Polybutadiene (butadiene rubber)	−100	5	P	P	F	Tire treads, shoes, conveyor belts
Acrylonitrile-butadiene copolymer (nitrile rubber)	−50 to −25		G	G	F	Fuel hoses gaskets, rollers
Isobutylene-isoprene copolymer (butyl rubber)	−70	−5	F	P	F	Tire liners, window strips
Ethylene-propylene monomer (EPM), ethylene-propylene-diene monomer (EPDM)	−55		F	P	F	Flexible seals, electrical insulation
Polychloroprene (neoprene)	−50	25	G	G	G	Hoses, belts, springs, gaskets
Polysulfide (Thiokol)	−50		F	E	F	Seals, gaskets, rocket propellants
Polydimethyl siloxane (silicone)	−125	−50	G	F	F	Seals, gaskets, surgical implants

Fluoroelastomer	−10		E	E	F	O-rings, seals, gaskets
Polyacrylate elastomer	−15 to −40		G	G	F	Hoses, belts, seals, coated fabrics
Polyethylene (chlorinated, chlorosulfonated)	−70		G	G	F	O-rings, seals, gaskets
Styrene-isoprene-styrene (SIS), styrene-butadiene-styrene (SBS) block copolymer	−60		P	P	F	Automotive parts, shoes, adhesives
EPDM-polypropylene blend	−50		F	P	F	Shoes, flexible covers

Vulcanization

Vulcanization is a chemical process by which the physical properties of natural or synthetic rubber are improved; finished rubber has higher tensile strength and resistance to swelling and abrasion, and is elastic over a greater range of temperatures. In its simplest form, vulcanization is brought about by heating rubber with sulfur.

The process was discovered in 1839 by the U.S. inventor Charles Goodyear, who also noted the important function of certain additional substances in the process. Such a material, called an accelerator (q.v.), causes vulcanization to proceed more rapidly or at lower temperatures. The reactions between rubber and sulfur are not fully understood, but in the product, the sulfur is not simply dissolved or dispersed in the rubber; it is chemically combined, mostly in the form of cross-links, or bridges, between the long-chain molecules.

In modern practice, temperatures of about 140°–180 °C are employed, and in addition to sulfur and accelerators, carbon black or zinc oxide is usually added, not merely as an extender, but to improve further the qualities of the rubber. Anti-oxidants are also commonly included to retard deterioration caused by oxygen and ozone. Certain synthetic rubbers are not vulcanized by sulfur but give satisfactory products upon similar treatment with metal oxides or organic peroxides.

Reinforcement

Employed as a pigment in Egyptian pottery dating to 4000 BC, carbon black is the pre-eminent reinforcing filler, able to impart a broad spectrum of properties to rubber compounds. There are over 40 grades of carbon black, with representative types listed in table. Carbon black consists of solid, colloidal (01 mm) entities called aggregates. Each aggregate is comprised of many primary particles fused together in a randomly arranged cluster, having a morphology akin to a bunch of grapes. The aggregate size and specific surface area are obviously important to rubber reinforcement, as are the number and arrangement of the particles within the aggregates. The latter govern the 'structure' of a given carbon black, which is a measure of the ratio of the effective volume of the aggregate to the sum of the primary particle volumes. Surface area can be measured by adsorption of a gas or an aqueous solution of surfactant (typically cetyltrimethylammonium bromide). The advantage of a fluid is that it is not absorbed in the angstrom–sized micropores of the aggregate that are likewise not accessed by polymer chains. Another common measure of surface area is the absorption of iodine, but although the method is very easy, the accuracy is poor, the results being affected by the surface chemistry of the filler. Structure is assessed from measurement of the internal void volume, usually by absorption of dibutyl phthalate, either on the carbon black as received ('DBPA') or after crushing and sieving the carbon particles ('CDBP' – crushed dibutyl phthalate). The surface area and structure of the particles determine how much rubber is 'immobilized' by the filler. The surface of carbon black is imperfect graphitic layers, with the carbon atoms at exposed edges containing C = O and C–OH groups, in the form of quinones, phenols, carboxyls, ketones, etc. Heating carbon black to high temperature (>2700 °C) in an inert atmosphere removes the oxygen and hydrogen; polymer chains do not react with or chemisorb to such 'graphitized' carbon black.

Table: Characteristics of typical carbon blacks.

ASTM type	Generic name	Particle size (nm)	Aggregate size (nm)	Surface area ($m^2 g^{-1}$)
N110	SAF	17±7	54±26	143
N220	ISAF	21±9	65±30	117
N330	HAF	31±13	86±44	80
N339	-	26±11	75±34	90
N351	-	31±14	89±47	75
N550	FEF	53±28	139±71	41
N660	GPF	63±36	145±74	34
N762	SRF	110±53	188±102	21
N990	MT	246±118	376±152	9

Colloidal silica is an alternative to carbon black, although typically the polarity difference between silica and common rubbers gives deficient reinforcing properties unless coupling agents are employed. Potential advantages of silica over carbon black include lower rolling resistance and reduced abrasive wear. Fumed silica is produced in a flame. Its aggregates tend to be less tightly clustered than those of carbon black, with silanol and nonpolar siloxane groups present on the surface. The main use for fumed silica is to reinforce silicone rubber, since its cost precludes more general application to rubbers.

The more common form of silica in the rubber industry is precipitated silica, formed by acidification of a sodium silicate solution (the same method used to form silica gel). Similar to carbon black, precipitated silica exists as aggregates, but unlike fumed silica, these aggregates tend to be more highly clustered, with some having the appearance of fragments of silica gel. The surface is covered with silanol groups, through which the aggregates bond to each other, as well absorb moisture. Achieving the reinforcing performance of carbon black with precipitated silica is problematic due to the different surface chemistry and morphology. Generally, in comparison to carbon black, silica has stronger particle-particle interactions, but weaker interactions with the rubber. Excellent properties have been obtained in tire compounds with precipitated silicas by improving the bonding to rubber, either by 'activating' the silica or by the addition of coupling agents to the compound.

A method that circumvents the problem of dispersing silica in rubber is by in situ precipitation, for example, via catalyzed hydrolysis of tetraethoxysilane. Small (<25 nm), irregularly shaped particles can be obtained, with both the particle size and degree of aggregation controlled by the precipitation and processing conditions. Although the method may have potential for industrial operations such as reactive extrusion processing, there is no industrial-scale application of this approach to date.

Other inorganic particulates, including kaolin clay and calcium carbonate, have found use in the industry, and as mentioned, zinc oxide was the first reinforcing filler for rubber. Since mineral fillers are cheaper than the polymer, they serve as low-cost extenders, while also increasing the modulus; however, they do not provide high degrees of compound reinforcement. As discussed above, the failure properties of mineral-filled compounds are poorer than rubber containing carbon black. Short fibers, including glass, cellulose, carbon, and aramid, are used to a limited extent in rubber, generally in combination with carbon black or silica. They increase the modulus and dimensional stability of rubber components, and given their large aspect ratio, can potentially yield anisotropic properties. Generally the fibers must be coated to facilitate dispersion and enable bonding to the elastomer.

A more recent development is the use of organo-modified, layered mineral silicates (clay). The material is relatively low cost and has been studied as a reinforcing filler in many polymers. Property improvements require intercalation, whereby polymer chains diffuse into the layer galleries, or exfoliation, in which there is separation of the silicate layers to yield nm-thick disks dispersed in the polymer. This 'nanoclay' is a two-dimensional filler, similar to graphene. The state of dispersion of the nanoclay can be deduced from X-ray measurements of the silicate d-spacing (lack of a diffraction peak indicating exfoliation). Although the performance enhancements and relatively low cost of organo-modified clays are attractive, inducing intercalation or exfoliation is a formidable problem, especially for non-polar polymers.

Dispersion is a general issue with the utilization of nano-particles. Carbon nanotubes, graphene, nano-diamond, as well as the modified silicates, have all shown the potential to yield enhanced properties, and a significant amount of research has been directed toward their application to rubbery materials. However, the high surface areas and large surface energies promote particle agglomeration. Laboratory studies rely on dispersion methods involving solvents, sonication, freeze-drying, chemical treatments, etc., none of which are especially amenable to economical scale-up. The key to exploiting nano-fillers in the rubber industry is overcoming this dispersion problem.

Mixing and Dispersion

Reinforcement requires good dispersion of the filler, which is accomplished on an industrial scale by mixing in an internal mixer or two-roll mill. This mixing is energy-intensive and has the potential to degrade the compound by chain-scission or premature curing. During mixing the filler aggregates become uniformly distributed (on a scale of tens to hundreds of microns), the polymer is incorporated into the void spaces of the agglomerated pellets, and ultimately (the most difficult step) the filler agglomerates are broken down into distinct aggregates. At the usual filler concentrations (volume fraction 10–20%), the dispersed aggregates have some contact with each other, even when well distributed. These contacts may increase after mixing, driven by enthalpic particle interactions; this is undesirable because it increases the mechanical hysteresis of the elastomer. Sufficient interaggregate contacts give rise to a filler network, which is manifested in an elevated dynamic modulus at low strains and, at least for carbon black, high electrical conductivity. Reagglomeration and network formation can be a particular issue with silica, causing hardening of the rubber prior to curing.

Dispersion of filler is important to minimize hysteresis and, since particles larger than the intrinsic flaw size, which is on the order of 10–30 mm, can act as defects very poor dispersion can affect failure properties. The complexity of the structure of carbon black, silica, etc. implies there are different degrees of dispersion. Conventional mechanical mixing does not fracture the aggregates, although some reduction in structure may occur.

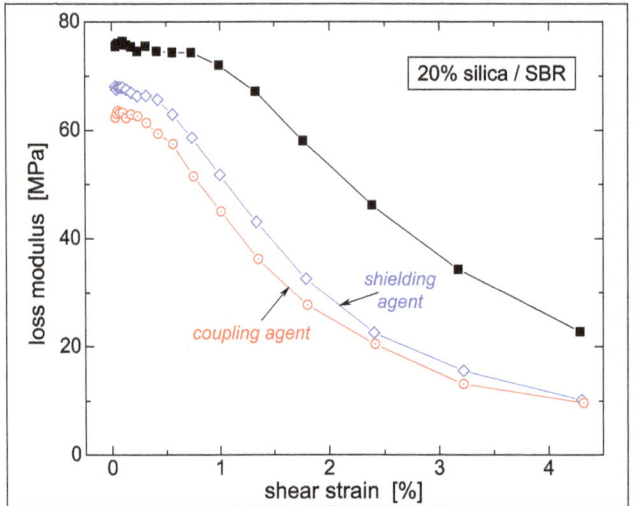

Dynamic loss modulus as a function of shear strain for SBR reinforced with 20% by volume silica: untreated (squares); with 6.5% by weight n-octyltriethoxysilane shielding agent (diamonds); or 4.7% by weight 3-mercaptopropyltrimethoxysilane coupling agent (circles). Suppression of agglomeration reduces the hysteresis.

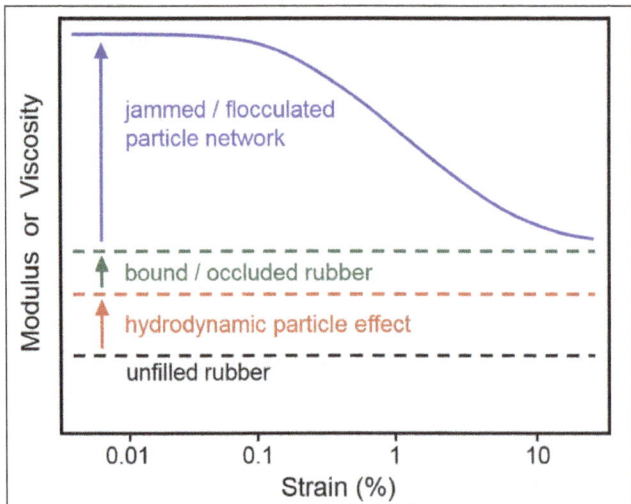

Schematic depicting the various contributions to the modulus and viscosity of rubber.

The term 'dispersion' refers to the breakup of agglomerates or pellets, leading to separation and uniform distribution of the aggregates within the polymer. Optical microscopy or surface roughness measurements reveal agglomerates more than roughly 2 mm in size. A finer level of dispersion is reflected in a reduced Payne effect and for carbon black compounds, higher electrical resistivity.

In addition to relying on adequate mechanical mixing to disperse the filler, polymers have been developed in which the chain ends are terminated with reactive moieties that bond to the carbon black to inhibit particle agglomeration. Also, coupling agents can be added to the rubber formulation for the same purpose. Figure shows dynamic mechanical loss measurements for a silica-reinforced styrene-butadiene copolymer. Reduction of the particle-particle interaction, and hence less energy dissipation, is achieved with either coupling or shielding agents, the latter reacting with the particle surface to suppress agglomeration.

Rheology and Modulus of Filled Rubber

There are various mechanisms underlying the effect of particulates on the deformation and flow of rubber. As illustrated in figure, these include a hydrodynamic effect that amplifies the strain in the polymer phase; the occlusion of polymer chains within the void structure of the particles, enhancing their effective concentration; and a filler network structure when the amount of particles exceeds their percolation threshold. We consider each of these contributions in turn.

Hydrodynamic Effect

Particulate fillers increase both the viscosity and elastic modulus of a polymer or rubber compound via a hydrodynamic effect – the inextensible particles cannot deform, so their strain is transferred to the polymer chains. This strain-amplification is described by an extension of the Einstein viscosity equation for dilute, spherical particles, to account for particle interactions at higher concentrations:

$$\eta = \eta_0(1 + 2.5\phi + 14.1\phi^2)$$
$$G' = G'_0(1 + 2.5\phi + 14.1\phi^2)$$

In equaton $\begin{aligned}\eta &= \eta_0(1 + 2.5\phi + 14.1\phi^2)\\ G' &= G'_0(1 + 2.5\phi + 14.1\phi^2)\end{aligned}$ ϕ is the volume fraction of the filler, η and G' are the respective viscosity and dynamic storage modulus, and the subscript zero refers to the gum (unfilled) compound. Equation $\begin{aligned}\eta &= \eta_0(1 + 2.5\phi + 14.1\phi^2)\\ G' &= G'_0(1 + 2.5\phi + 14.1\phi^2)\end{aligned}$ is applicable, for example, to rubber filled with N990, a low surface area, low structure carbon black. However, for more reinforcing fillers, a significant amount of rubber becomes occluded within the void space of the aggregates. This occluded polymer, which amplifies the hydrodynamic interaction, can be accounted for by including it in ϕ; that is, the magnitude of ϕ represents an effective volume fraction of filler. Non-spherical particles require a further modification of equation $\begin{aligned}\eta &= \eta_0(1 + 2.5\phi + 14.1\phi^2)\\ G' &= G'_0(1 + 2.5\phi + 14.1\phi^2)\end{aligned}$; a common expression for rodlike filler particles is the Guth– Gold equation.

$$\eta = \eta_0(1 + 0.67f\phi + 1.62f^2\phi^2)$$

Which can also applied to the dynamic modulus. The aspect ratio of the rods is f; but for particles that are not rod-shaped, this parameter can be adjusted to fit the experimental data, yielding an effective shape factor (which underestimates the actual value). In figure (lower panel) are the viscosities for a polytetramethylene oxide with varying concentrations of multiwall carbon nanotubes. Fitting equation $\eta = \eta_0(1 + 0.67f\phi + 1.62f^2\phi^2)$ to these data yields f=21 for the particle aspect ratio. This is much less than the value for a single nanotube, indicating agglomeration. This conclusion can be tested by analysis of the percolation threshold of the material, since the overlap concentration for the filler also depends on f (eqn $\phi^* = \dfrac{\pi}{4}f^{-2}$).

Rubber–Filler Interaction

Polymer chains will attach to reinforcing filler particles, especially carbon black, via chemisorption, or even covalent bonding if the filler has reactive sites. The term 'bound rubber' refers to chains bonded to the particles sufficiently strongly to resist dissolution in a solvent.

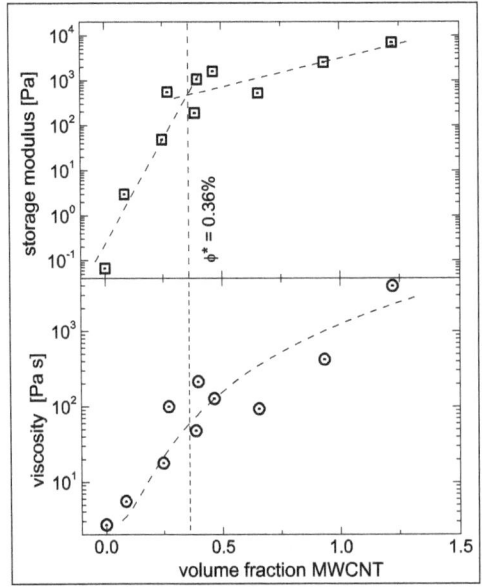

(Lower) Viscosity of PTMO as a function of filler content; the fitted line is the Guth–Gold equation with f=21. (Upper) Storage modulus of the same compound, with the change in slope at a value of ϕ for which the percolation equation (eqn $\phi^* = \dfrac{\pi}{4}f^{-2}$) gives f=16.

The amount of bound rubber generally correlates with the occluded rubber, as governed by the particle's surface area and structure. Precipitated silica gives considerably less bound rubber than a carbon black of similar surface area, and graphitized carbon black, in which surface imperfections and layer edges have been reduced by heating, yields negligible bound rubber.

The tethering of chains to the particle surface of the chains is expected to cause reduced mobility within the bound layer. It is for this reason that ϕ in equation

$$\eta = \eta_0(1 + 2.5\phi + 14.1\phi^2)$$
$$G' = G'_0(1 + 2.5\phi + 14.1\phi^2).$$

underestimates the hydrodynamic effect, unless occluded rubber is included. However, this restriction on the mobility is limited. The occluded layer does not lose configurational freedom or become glassy. The local segmental dynamics, encompassing correlated rotation of several backbone bonds, is usually unaffected by the presence of reinforcing filler.

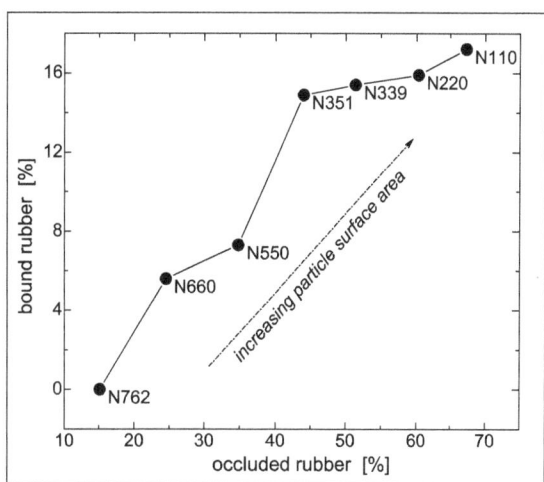

Quantity of rubber bonded to the filler versus the amount of rubber occluded in the particle structure for a 1,4-polybutadiene with 18% by volume carbon black of the indicated type.

Filler–Filler Interaction (Particle Network Formation)

The third mechanism for modulus enhancement of reinforced rubber is formation of a network of filler particles. This is the largest contribution to the modulus in figure, but requires a filler concentration above the threshold for percolation (macroscopically extended interparticle contacts). For randomly oriented particles, this overlap concentration is given by:

$$\phi^* = \frac{\pi}{4} f^{-2}$$

The dependence of the storage modulus on MWCNT concentration for the polymer in figure (upper panel) shows a change in slope at $\phi = 0.36\%$. Identifying this with the overlap concentration, eqn $\phi^* = \frac{\pi}{4} f^{-2}$ yields f=15, consistent with the aspect ratio determined from fitting eqn

$$\eta = \eta_0 (1 + 0.67 f\phi + 1.62 f^2 \phi^2)$$ to the viscosity of the material.

The percolated filler network is disrupted by deformation (Payne effect); displacement of the particles with increasing dynamic strain amplitude causes a substantial decrease in the storage modulus and a peak in the loss modulus. This behavior is shown in figure for natural rubber containing various levels of N110 carbon black. Unfilled compounds do not exhibit this strain dependence of their dynamic properties. While displacement of the interaggregate contacts is the main origin of the Payne effect, detachment of bound polymer and disentanglement of chains can also be contributing factors. Although the strain associated with breakup of the aggregate network varies with filler type and other variables, the required strain energy, calculated as the product of the stress and strain amplitudes, is constant for a given polymer. For carbon black and silica, this critical strain energy is in the range from 2–4 kJ m^{-3}.

Mechanical Hysteresis

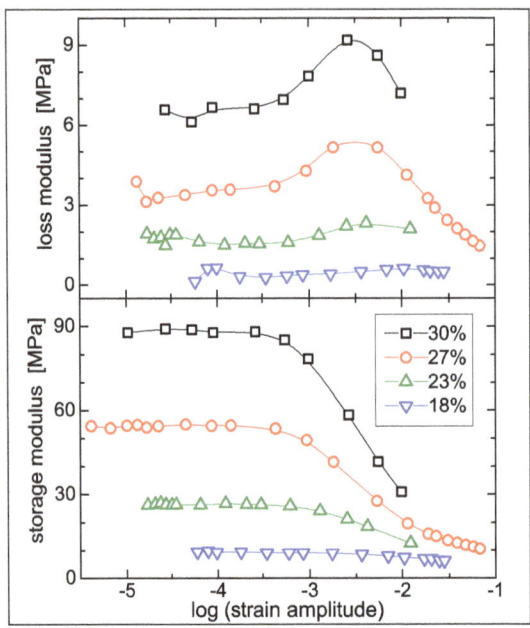

Payne effect in natural rubber containing the indicated amount (by volume) of N110 (SAF) carbon black.

Mechanical hysteresis, which reduces efficiency and is responsible for heat buildup, is an important property of rubber. A prominent example is the braking performance and rolling resistance of tires, both related to mechanical energy losses. Hysteresis is observed in rubber subjected to either dynamic deformation or reversing strain histories. Since the magnitude of the strain is usually very different for these two types of deformation, the mechanisms contributing to the hysteresis can be different.

Dynamic Properties

Under normal dynamic strain conditions, a filler-reinforced elastomer exhibits more hysteresis than its unfilled counterpart; the primary mechanism is the breakup and reformation of filler interparticle bonds. The energy loss depends not only on the mechanical properties of the compound, but is also affected by the nature of the deformation. In terms of the storage, G' , and loss, G'' , moduli, the energy loss for various deformation conditions is given by:

$$= \pi \sigma_0^2 (G'' / [G'^2 + G''^2]) \quad \text{(Constant stress)}$$
$$E_{loss} = \pi \varepsilon_0^2 G'' \quad\quad\quad\quad\quad \text{(Constant strain)}$$
$$= \pi \varepsilon_0 \sigma_0 G'' / G' \quad\quad\quad \text{(Constant energy)}$$

Where σ_0 and ε_0 are the respective amplitudes of the stress and strain. Complex deformations, such as that of a rolling tire, conform to none of these three conditions; however, the strain can be characterized in terms of a deformation index, n, which equals 0, 1, or 2 for constant strain, constant energy, or constant stress, respectively. The energy loss is related to n according to:

$$E_{loss} = zG'' / \left(G'^2 + G''^2\right)^{n/2}$$

in which z is a constant. From equation $E_{loss} = zG'' / \left(G'^2 + G''^2 \right)^{n/2}$ and the measured energy loss, the value of n can be determined, which via equation:

$$E_{loss} = \begin{cases} \pi\sigma_0^2(G''/[G'^2 + G''^2]) & \text{(Constant stress)} \\ \pi\varepsilon_0^2 G'' & \text{(Constant strain)} \\ \pi\varepsilon_0\sigma_0 G''/G' & \text{(Constant energy)} \end{cases}$$

reveals the dynamic properties governing hysteresis for the given deformation.

Mechanical energy loss in a tire, known as rolling resistance, contributes to fuel consumption (roughly 10% for cars; at least twofold higher for trucks). However, a perfectly elastic tire would not dissipate sufficient energy to stop the vehicle during skidding. To overcome the contradictory requirements of low rolling resistance but adequate braking performance, tire treads have dissipation mechanisms only active at the high strain rates encountered during wet skidding. These generally rely on a sufficiently high glass transition temperature (>230 K), so that the polymer segmental dynamics are on the order of 10^5 Hz or higher at ambient temperature. The energy dissipation from the Payne effect is essentially independent of strain rate and contributes under all conditions. Thus, mixing procedures are designed to minimize the interparticle contacts and consequent network formation, in order to reduce rolling resistance without adverse effect on wet skid resistance.

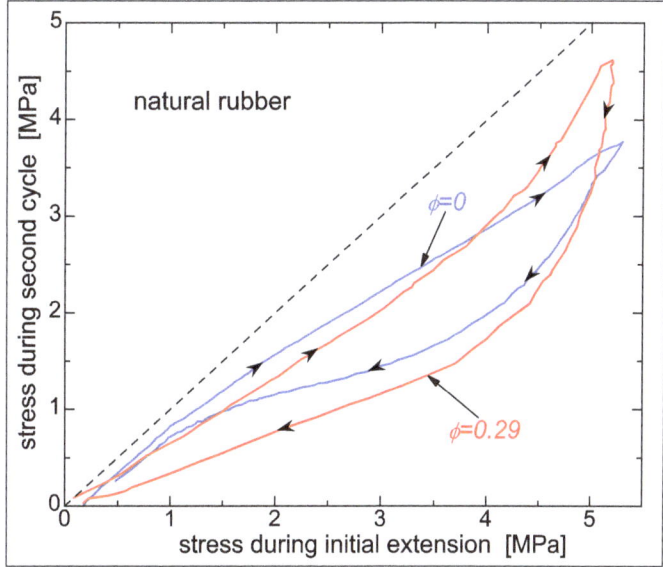

Stress measured during extension–retraction of natural rubber unfilled (blue line) and containing 29% by volume (80 phr) N330 carbon black (red line). The abscissa is the stress for the first extension cycle and the ordinate is the stress for subsequent cycles. The dashed line corresponds to zero hysteresis.

Strain Softening and the Mullins Effect

Uniaxial mechanical testing at a nominally constant strain rate is common in the characterization of rubber, and the effects of reinforcement are manifested in such measurements. The Young's modulus is higher due to the hydrodynamic effect and to the filler network, present. However, if the testing is carried out below the point of failure, the behavior during retraction can be observed, and this invariably exhibits large mechanical hysteresis. Known as the Mullins effect, this hysteresis is commonly assumed to be due to the filler. Actually, such softening is an inevitable consequence of polymer viscoelasticity that can be reproduced with simple spring–dashpot models. The basic mechanism is the retarded response of the polymer chains, which reduces the perturbation during retraction. The result is lower retraction stresses. Mullins softening can be reproduced in mathematical models by the use of a damping function.

In addition to this viscoelastic 'softening,' strain crystallization can cause large hysteresis during extension–retraction of rubber. Direct evidence that the Mullins effect is unrelated to filler reinforcement is the equivalent hysteresis of gum and filled rubber, when compared at similar stress levels.

Failure Properties of Filled Rubber

Particulate fillers improve the strength and fatigue resistance of rubber. This is illustrated in figure, showing substantial increases in tensile strength, even for a filler such as calcium carbonate that provides minimal actual reinforcement. Similar increases in resistance to tearing and crack growth are shown in tables These improvements may seem surprising, since the hydrodynamic

effect (eqs $\begin{aligned} \eta &= \eta_0(1+2.5\phi + 14.1\phi^2) \\ G' &= G'_0(1+2.5\phi + 14.1\phi^2) \end{aligned}$ and $\eta = \eta_0(1 + 0.67f\phi + 1.62f^2\phi^2)$)) amplifies the strain

of the polymer chains, and thus also amplify the local stresses. Evidently there are mechanisms that toughen the material, at least locally, to yield a net enhancement in performance. Although the topic remains to be fully understood, processes that contributed to better failure properties include strain energy dissipation from filler–polymer detachment; mitigation of stress concentration by load transfer through the particle when a chain breaks; and crack growth deviation around filler particles. The strength of rubber increases with filler content, attaining a broad maximum at high concentrations.

There is a prevailing notion that the failure properties of strain-crystallizing rubbers are less affected by fillers. This is based on the idea that at the crack front, where stresses are highest and reinforcement most required, crystallites form that blunt the crack tip and toughen the material locally; thus, the need for filler reinforcement is diminished. Interestingly, although it is true that the failure properties of strain-crystallizing elastomers are generally superior to those of their amorphous counterparts, the tear, and crack growth resistance of the former still improve significantly when particulate fillers are incorporated in the compound. This is seen in the tear strength data in table, comparing natural rubber and a non-crystallizing SBR, and table, comparing cut growth rates for several strain-crystallizing and amorphous rubbers. Note in table, with the exception of polychloroprene, property enhancement is greater for the higher structure filler (N330) than for the less reinforcing, thermal black (N990).

Tire wear, especially during hard braking or skidding, involves high temperatures that may preclude strain crystallization. Also, crystallization, which is not instantaneous, cannot occur at very high strain rates (exceeding c. 20 s^{-1}). In these situations, filler reinforcement becomes mandatory, even for crystallizing rubbers, in order to achieve optimal performance.

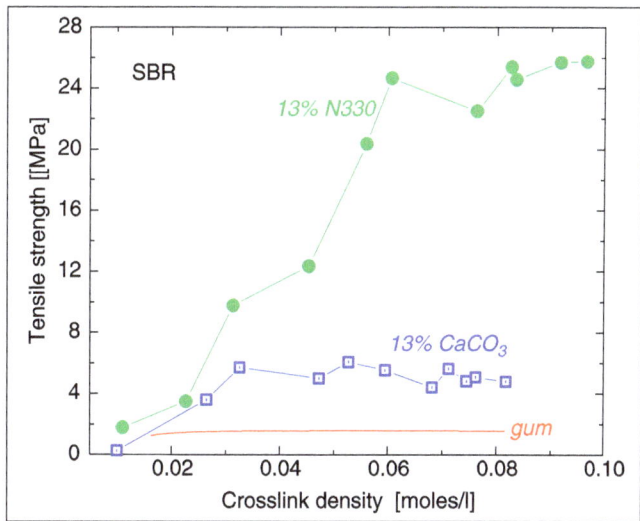

Tensile strength of SBR unfilled (line), with 0.13 volume fraction (45 phr) calcium carbonate (squares), and with 0.13 volume fraction (30 phr) HAF carbon black, as a function of crosslink density.

Table: Effect of reinforcing filler of tear strength.

Rubber	Filler	Tensile strength (MPa)	Tear test	
			Pure shear (kJ m^{-2})	Trouser tear (kJ m^{-2})
NR	Gum	28.3	6	12
	N330	28.5	12	50
SBR	Gum	2.3	1.4	3.2
	N330[a]	28.7	7	6

Table: Effect of filler on cut growth rates.

Rubber	Strain crystallize?	Carbon black[a]	Cut growth rate × 10^{-5} (mm cycle^{-1})[b]	Relative cut growth rate
NR	Yes	Gum	5.60	1
		N990	2.03	36%
		N330	0.182	0.3%
CR	Yes	Gum	25.5	1
		N990	2.71	11%
		N330	11.1	43%
IIR	Yes	Gum	97.9	1
		N990	8.89	9%
		N330	0.137	0.1%
PB	No	Gum	247	1
		N990	163	66%
		N330	1.13	0.5%
NBR	No	Gum	9.30	1
		N990	3.11	33%
		N330	0.87	9%
SBR	No	Gum	146	1
		N990	14.3	9.8%
		N330	12.1	0.6%

High surface area carbon blacks are considered essential for good wear. However, wear mechanisms are complicated, and since the structure and quantity of the filler affects various compound properties, there are no general principles applicable to selection of an optimal filler for abrasion resistance.

The majority of applications of rubber, certainly those involving mechanical properties, require reinforcement, with carbon black and precipitated silica the dominant fillers. Reinforcing fillers increase the viscosity and modulus by a hydrodynamic effect, augmented by rubber occluded within the filler. At low strain amplitude, there can be an additional contribution from an interparticle network; breakdown and re-formation of this network during cyclic straining contributes to the hysteresis of rubber compounds. The ultimate properties (tensile strength, fatigue life, abrasive wear, etc.) are an important aspect of reinforcement, although the magnitude of the improvements and the relevant mechanisms are not always well understood. This chapter was limited mainly to conventional fillers. Nanofillers show great promise and their commercial potential is significant. However, the dispersion problem, encountered in the use of all reinforcing fillers, remains a major obstacle to utilization of nanoparticles in the rubber industry.

Elastomer Properties and Compounding

Elastomers are amorphous polymers with considerable segmental motion. Their general molecular form has been likened to a "spaghetti and meatball" structure, where the meatballs signify cross-links between the exible polymer chains, which are like spaghetti strands. Each polymer chain is made up of many monomer subunits, and each monomer is usually made of carbon, hydrogen, and oxygen atoms, and occasionally silicon atoms.

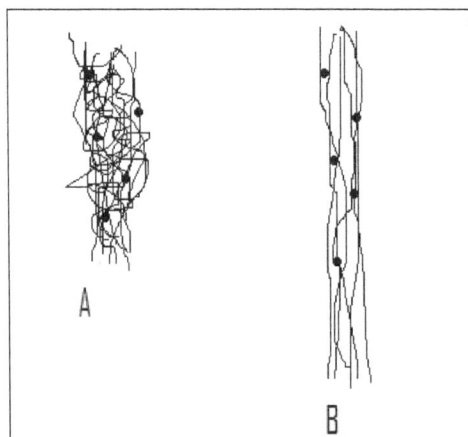

A is a schematic drawing of an unstressed polymer, and B is the same polymer under stress. When the stress is removed, B returns to the A conguration. (The dots represent cross-links).

Most elastomers are thermosets—that is, they require curing (by heat, chemical reaction, or irradiation). In the curing process, the long polymer chains become cross-linked by covalent bonds, the material becomes stronger, and it cannot be remelted and remolded. Some elastomers are thermoplastic, melting to a liquid state when heated and turning brittle when cooled suciently. In thermoplastic elastomers, the polymer chains are cross-linked by weaker bonds, such as hydrogen bonds or dipole-dipole interactions.

The elasticity is derived from the ability of the long chains to recongure themselves to distribute an applied stress. Covalent cross-linkages, in particular, ensure that the elastomer will return to its original conguration when the stress is removed.

The temperature of the polymer also affects its elasticity. Elastomers that have been cooled to a glassy or crystalline phase will have less mobile chains, and consequently less elasticity, than those manipoulated at temperatures higher than the glass transition temperature of the polymer. At ambient temperatures, rubbers are thus relatively soft (Young's modulus of about 3 MPa) and deformable.

References

- Plastic, science: britannica.com, Retrieved 1 March, 2019

- Plastic-molding, custom-rotational-molding: fibertechinc.net, Retrieved 12 May, 2019

- Extrusion-plastic, plastics-rubber: thomasnet.com, Retrieved 4 January, 2019

- Methods-manufacturing-plastic-72181: chron.com, Retrieved 15 July, 2019

- The-processing-and-fabrication-of-plastics, plastic, science: britannica.com, Retrieved 3 February, 2019

- The-fibers: fibrenamics.com, Retrieved 31 January, 2019

- Fiber-Spinning: umass.edu, Retrieved 20 April, 2019

- Elastomer, science: britannica.com, Retrieved 26 February, 2019

- Vulcanization, technology: britannica.com, Retrieved 14 June, 2019

- What-is-elastomer-and-properties-of-elastomer: semesters.in, Retrieved 2 August, 2019

4

Understanding Copolymerization

The polymer which is created when more than one type of polymers are connected in the same polymer chain is known as copolymer. This chapter has been carefully written to provide an easy understanding of the varied facets of copolymerization as well as copolymerization reactions and kinetics.

Copolymers

A copolymer is a polymer that is made up of two or more monomer species. Many commercially important polymers are copolymers. Examples include polyethylene-vinyl acetate (PEVA), nitrile rubber, and acrylonitrile butadiene styrene (ABS). The process in which a copolymer is formed from multiple species of monomers is known as copolymerization. It is often used to improve or modify certain properties of plastics.

A homopolymer is a polymer that is made up of only one type of monomer unit. The difference in the constitution of a copolymer and a homopolymer is illustrated below.

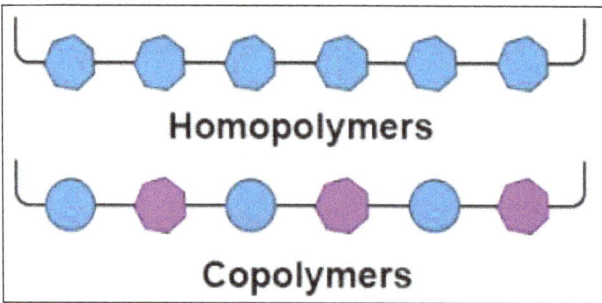

Copolymers are categorized based on their structures. Those containing a single chain are known as linear copolymers whereas those containing polymeric side chains are called branched copolymers.

Classification of Copolymers

What are the Different Types of Linear Copolymers?

Linear copolymers can be further classified into several categories such as alternating and statistical copolymers. This classification is done based on the arrangement of the monomers on the main chain.

Block Copolymers

- When more than one homopolymer units are linked together via covalent bonds, the resulting single-chain macromolecule is called a block copolymer.

- The intermediate unit at which the two homopolymer chains are linked is called a junction block.

- A diblock copolymer contains two homopolymer blocks whereas a triblock copolymer contains three distinct blocks of homopolymers.

- An example of such a polymer is acrylonitrile butadiene styrene, commonly referred to as SBS rubber.

- An illustration describing the structure of a block copolymer which is made up of the monomers 'A' and 'B' is provided below.

Statistical Copolymer

- Statistical copolymers are the polymers in which two or more monomers are arranged in a sequence that follows some statistical rule.

- Should the mole fraction of a monomer be equal to the probability of finding a residue of that monomer at any point in the chain, the entire polymer is then known as a random polymer.

- These polymers are generally synthesized via the free radical polymerization method.

- An example of a statistical polymer is the rubber made from the copolymers of styrene and butadiene.

- An illustration describing the structure of a statistical copolymer is provided below.

Alternating Copolymers

- Alternating copolymers contain a single main chain with alternating monomers.

- The formula of an alternating copolymer made up of monomers A and B can be generalized to $(-A-B-)_n$.

- Nylon 6,6 is an example of an alternating copolymer, consisting of alternating units of hexamethylene diamine and adipic acid.

- An illustration describing the general structure of these polymers is provided below.

Periodic Copolymers

These polymers feature a repeating sequence in which the monomers are arranged in a single chain. An illustration of the structure of a periodic copolymer made up of monomers A and B is provided below.

$$\left\{ A - B - A - B - B - A - A - A \right\}_n$$

Gradient and Stereoblock Copolymers

The single-chain copolymers in which the composition of monomers gradually changes along the main chain are called gradient copolymers. If the tacticity of the monomers varies with different blocks or units in the polymer, the macromolecule is known as a stereoblock copolymer.

What is a Branched Copolymer?

As the name suggests, a branched copolymer is a polymer in which the monomers form a branched structure. Some important types of branched copolymers include star, comb, grafted, and brush copolymers.

A star copolymer contains several polymeric chains that are attached to the same central core.

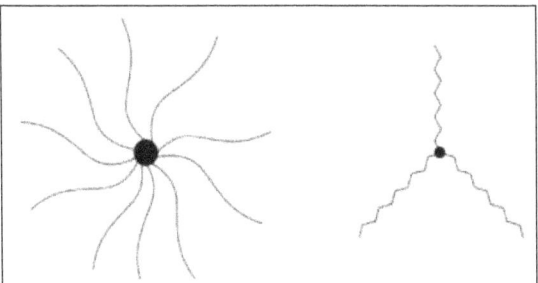

The structures of different types of star-shaped copolymers are illustrated above. They consist of a multifunctional centre to which three or more polymer chains are attached.

Graft Copolymers

Branched copolymers featuring differently structured main chains and side chains are known as graft copolymers. An illustration detailing the structure of a graft copolymer made up of monomers A and B is provided below.

The main chain or the side chains of these polymers can be copolymers or homopolymers. High impact polystyrene is an important example of a graft copolymer. They can be synthesized from free radical polymerization.

Copolymerization Reactions

It isn't difficult to form addition polymers from monomers containing C=C double bonds; many of these compounds polymerize spontaneously unless polymerization is actively inhibited. One of the problems with early techniques for refining gasoline, for example, was the polymerization of alkene components when the gasoline was stored. Even with modern gasolines, deposits of "gunk" can form when a car or motorcycle is stored for extended periods of time without draining the carburetors.

The simplest way to catalyze the polymerization reaction that leads to an addition polymer is to add a source of a free radical to the monomer. The term free radical is used to describe a family of very reactive, short-lived components of a reaction that contain one or more unpaired electrons. In the presence of a free radical, addition polymers form by a chain-reaction mechanism that contains chain-initiation, chain-propagation, and chain- termination steps.

Chain Initiation

A source of free radicals is needed to initiate the chain reaction. These free radicals are usually produced by decomposing a peroxide such as di-tert-butyl peroxide or benzoyl peroxide, shown below. In the presence of either heat or light, these peroxides decompose to form a pair of free radicals that contain an unpaired electron.

Chain Propagation

The free radical produced in the chain-initiation step adds to an alkene to form a new free radical.

The product of this reaction can then add additional monomers in a chain reaction.

Chain Termination

Whenever pairs of radicals combine to form a covalent bond, the chain reactions carried by these radicals are terminated.

$$CH_3-\underset{\underset{CH_3}{|}}{\overset{\overset{CH_3}{|}}{C}}-\ddot{\ddot{O}}-(CH_2CH_2)_n\bullet \quad + \quad \bullet(CH_2CH_2)_m-\ddot{\ddot{O}}-\underset{\underset{CH_3}{|}}{\overset{\overset{CH_3}{|}}{C}}-CH_3 \quad\longrightarrow\quad$$

$$CH_3-\underset{\underset{CH_3}{|}}{\overset{\overset{CH_3}{|}}{C}}-\ddot{\ddot{O}}-(CH_2CH_2)_{n+m}-\ddot{\ddot{O}}-\underset{\underset{CH_3}{|}}{\overset{\overset{CH_3}{|}}{C}}-CH_3$$

Formation of Branched Polymers

At first glance we might expect the product of the free-radical polymerization of ethylene to be a straight-chain polymer. As the chain grows, however, it begins to fold back on itself. This allows an intramolecular reaction to occur in which the site at which polymerization occurs is transferred from the end of the chain to a carbon atom along the backbone.

When this happens, branches are introduced onto the polymer chain. Free-radical polymerization of ethylene produces a polymer that contains branches on between 1 and 5% of the carbon atoms. Of these branches, 10% contain two carbon atoms, 50% contain four carbon atoms, and 40% are longer side chains.

Anionic Polymerization

Addition polymers can also be made by chain reactions that proceed through intermediates that carry either a negative or positive charge.

When the chain reaction is initiated and carried by negatively charged intermediates, the reaction is known as anionic polymerization. Like free-radical polymerizations, these chain reactions take place via chain-initiation, chain-propagation, and chain-termination steps.

The reaction is initiated by a Grignard reagent or alkyllithium reagent, which can be thought of a source of a negatively charged CH_3^- or $CH_3CH_2^-$ ion.

The CH_3^- or $CH_3CH_2^-$ ion from one of these metal alkyls can attack an alkene to form a carbon-carbon bond.

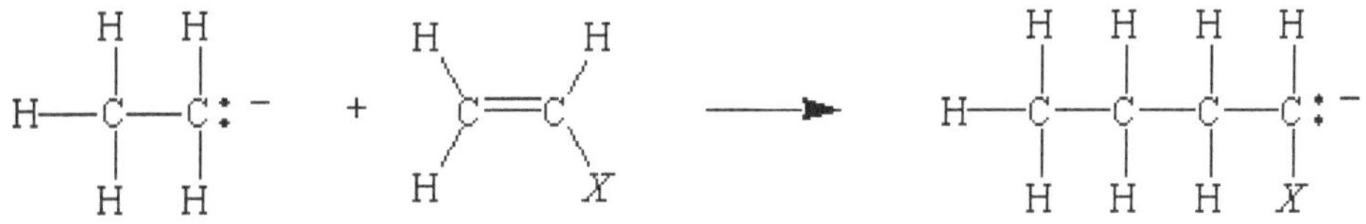

The product of this chain-initiation reaction is a new carbanion that can attack another alkene in a chain-propagation step.

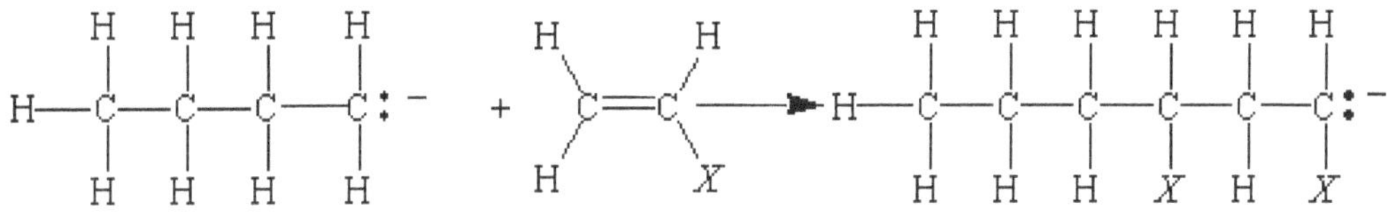

The chain reaction is terminated when the carbanion reacts with traces of water in the solvent in which the reaction is run.

$$CH_3CH_2(CH_2CH)_nCH_2CH_2^- \;+\; H_2O \longrightarrow CH_3CH_2(CH_2CH)_nCH_2CH_2 \;+\; OH^-$$

(with X substituents on the indicated carbons)

Cationic Polymerization

The intermediate that carries the chain reaction during polymerization can also be a positive ion, or cation. In this case, the cationic polymerization reaction is initiated by adding a strong acid to an alkene to form a carbocation.

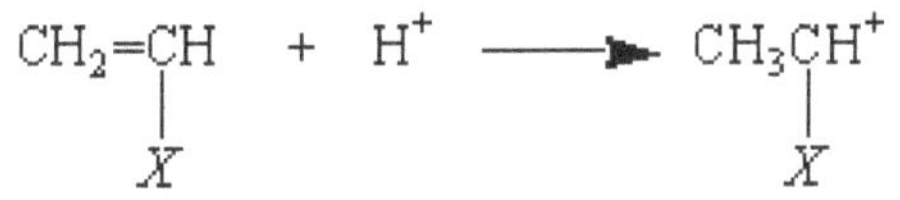

$$CH_2=CH \;+\; H^+ \longrightarrow CH_3CH^+$$

(with X substituent)

The ion produced in this reaction adds monomers to produce a growing polymer chain.

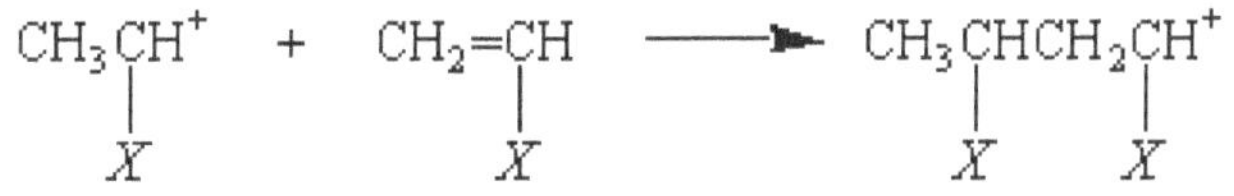

$$CH_3CH^+ \;+\; CH_2=CH \longrightarrow CH_3CHCH_2CH^+$$

(with X substituents)

The chain reaction is terminated when the carbonium ion reacts with water that contaminates the solvent in which the polymerization is run.

Advantages of Free-radical versus Ionic Polymerization

The initiation step of ionic polymerization reactions has a much smaller activation energy than the equivalent step for free-radical polymerizations. As a result, ionic polymerization reactions are relatively insensitive to temperature, and can be run at temperatures as low as -70 °C. Ionic polymerization therefore tends to produce a more regular polymer, with less branching along the backbone, and more controlled tacticity.

Because the intermediates involved in ionic polymerization reactions can't combine with one another, chain termination only occurs when the growing chain reacts with impurities or reagents that can be specifically added to control the rate of chain growth. It is therefore easier to control the average molecular weight of the product of ionic polymerization reactions.

Ionic polymerizations are more difficult to carry out on an industrial scale than free-radical polymerizations. Ionic polymerization is therefore only used for monomers that don't polymerize by the free-radical mechanism or to prepare polymers with a regular structure.

Coordination Polymerization

In 1963 Karl Ziegler and Giulio Natta received the Nobel prize in chemistry for their discovery of coordination compound catalysts for addition polymerization reactions. These Ziegler-Natta catalysts provide the opportunity to control both the linearity and tacticity of the polymer.

Free-radical polymerization of ethylene produces a low-density, branched polymer with side chains of one to five carbon atoms on up to 3% of the atoms along the polymer chain. Ziegler-Natta catalysts produce a more linear polymer, which is more rigid, with a higher density and a higher tensile strength. Polypropylene produced by free-radical reactions, for example, is a soft, rubbery, atactic polymer with no commercial value. Ziegler-Natta catalysts provide an isotactic polypropylene, which is harder, tougher, and more crystalline.

A typical Ziegler-Natta catalyst can be produced by mixing solutions of titanium(IV) chloride ($TiCl_4$) and triethylaluminum [$Al(CH_2CH_3)_3$] dissolved in a hydrocarbon solvent from which both oxygen and water have been rigorously excluded. The product of this reaction is an insoluble olive-colored complex in which the titanium has been reduced to the Ti(III) oxidation state.

The catalyst formed in this reaction can be described as coordinately unsaturated because there is an open coordination site on the titanium atom. This allows an alkene to act as a Lewis base toward the titanium atom, donating a pair of electrons to form a transition-metal complex.

The alkene is then inserted into a Ti-CH_2CH_3 bond to form a growing polymer chain and a site at which another alkene can bond.

Thus, the titanium atom provides a template on which a linear polymer with carefully controlled stereochemistry can grow.

Kinetics of Copolymerization

(a) Kinetics: Free living copolymerization proceeds by the attachment and detachment of different monomers $m_j \in \{1,2,...,M\}$ to macromolecular chains:

$$m_1 m_2 ... m_{l-1} + m_l \rightleftharpoons m_1 m_2 ... m_{l-1} m_l$$

The process is supposed to go on without termination at one end of the chain. The solution is large enough so that the monomeric concentrations remain constant and the process can reach a regime of steady growth. Copolymerization can be promoted by a catalyst located at the end of the chain. Moreover, an external force f may be exerted on the catalyst. The control parameters are thus the monomeric concentrations $\{c_m\}_{m=1}^{M}$ and the possible external force f. If the solution is dilute, the copolymers do not interact with each other so that their concentrations are proportional to the probabilities that a single copolymer has the sequence $m_1 m_2 \cdots m_l$ at the time t:

$$P_t(m_1...m_l) = \frac{V}{N} c_t(m_1 m_2 ... m_l)$$

In a volume V containing N copolymers. The time evolution of these probabilities is ruled by the kinetic equations:

$$\frac{\mathrm{d}}{\mathrm{d}t} P_t(m_1...m_{l-1}) = W_{+m_1,l} P_t(M_1...m_{l-1}) + \sum_{m_{l+1}=l}^{M} W - m_{l+1} l + 1 P_t(m_1...m_{l-1}m_l m_{l+1})$$

$$- \left(W - m_{1,l} + \sum_{m_{l+1}=1}^{M} W + m_{l+1,} l + 1 \right) P_t(M_1...m_{l-1}m_l)$$

Where the first gain term describes the attachment of the monomer m_l to $m_1 \cdots m_{l-1}$, the second terms the detachment of the monomeric units m_{l+1} from $m_1 \cdots m_{l-1} m_l m_{l+1}$, and the loss terms the events occurring to the chain $m_1 \cdots m_{l-1} m_l$ itself. The rates of attachment and detachment $W_{\pm m l'} l$ depend in general on the monomer m_l that is attached or detached, possibly on the configuration of the macromolecule of length l, as well as on the external control parameters.

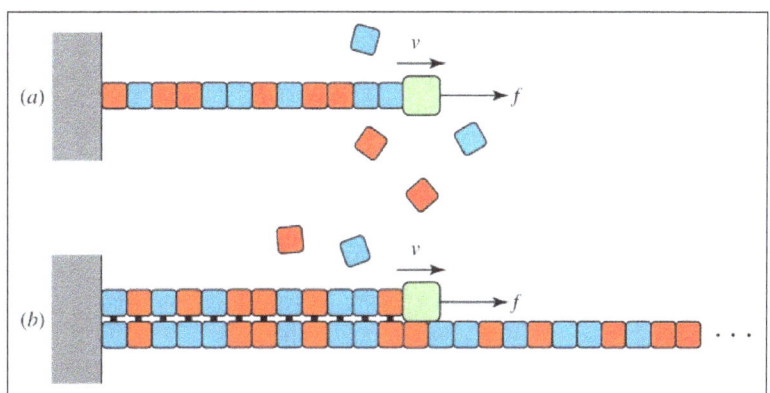

Schematic representations of (a) free living copolymerization and (b) template-directed living copolymerization promoted by a catalyst located at the growing end of the copolymer. f denotes the possible external force exerted on the catalyst and v is the mean growth velocity.

In the regime of steady growth, the growing chain has sequences characterized by a stationary probability distribution $\mu_l(m_1 \cdots m_l)$ so that the probabilities factorize as:

$$P_t(m_1...m_l) \simeq p_t(l)\mu_l(m_1...m_l)$$

where $p_t(l)$ is the probability that the copolymer has the length l at time t. After a long enough time, this latter becomes a Gaussian distribution of mean length $\langle l \rangle_t \simeq vt$ with a constant growth velocity v and a variance also increasing linearly in time.

If the rates depend on k monomeric units behind the last one $l \leftrightarrow m_{l-1}...m_{l-k}$, it is known that the growing copolymer is a kth-order Markov chain so that the sequence probability distribution itself factorizes as:

$$\mu_l(m_1...m_l) = \prod_{j=1}^{l-k} \mu(m_j \mid m_{j+1}...m_{j+k})\mu(m_{l-k+1}...m_l) \quad \text{if } l > k,$$

in terms of the conditional probabilities $\mu(m_j \mid m_{j+1} \cdots m_{j+k})$ of the Markov chain and the tip probabilities $\mu(m_{l-k+1} \cdots m_l)$.

Analytical methods have been developed to determine these probabilities as well as the growth velocity.

(b) Thermodynamics: In steady growth regimes, the entropy production of processes ruled by the kinetic equation:

$$\frac{d}{dt} P_t(m_1...m_{l-1}) = W_{+m_1,l} P_t(M_1...m_{l-1}) + \sum_{m_{l+1}=l}^{M} W - m_{l+1}, l+1 P_t(m_1...m_{l-1}m_l m_{l+1})$$

$$- \left(W - m_{1,l} + \sum_{m_{l+1}=1}^{M} W + m_{l+1}, l+1 \right) P_t(M_1...m_{l-1}m_l),$$

can generally be expressed as:

$$\frac{1}{k\text{B}} \frac{di\,S}{dt} = vA = v(\in + D) \geq 0,$$

in terms of Boltzmann's constant k_B, the growth velocity v and the affinity A, which is the sum of the free-energy driving force:

$$\in \equiv \lim_{l \to \infty} \frac{1}{l} \sum_{m_1...m_l,l} \mu_l(m_1...m_l) \ln \frac{W + m_l, l}{W - m_l, l}$$

and the Shannon disorder per monomeric unit in the growing sequences:

$$D \equiv \lim_{l \to \infty} -\frac{1}{l} \sum_{m_1...m_l} \mu_l(m_1...m_l) \ln \mu_l(m_1...m_l)$$

This latter is positive for a copolymer and zero for a pure polymer.

Because the entropy production is always non-negative by the second law of thermodynamics, the affinity must be positive in a growth regime. Therefore, the growth is possible either in a favourable free-energy landscape if the free-energy driving force is positive, or in an adverse free-energy landscape if $-D<\epsilon\leq0$. In the latter case, the growth is driven by the entropic effect of sequence disorder. At thermodynamic equilibrium, the growth velocity vanishes together with the affinity so that the free-energy driving force is equal to minus the sequence disorder: $\epsilon_{eq}=-D_{eq}$.

(c) The case of Bernoulli chains: If the rates only depend on the monomeric unit that is attached or detached $W_{\pm ml},l=w_{\pm ml}$, the process yields Bernoulli chains (corresponding to $k=0$). Accordingly, the probability distribution factorizes into the probabilities $\mu(m_j)$ to find the monomeric unit m_j anywhere in the sequence. If the growth velocity is equal to v, the net incorporation rate of the monomeric unit m is given by its attachment rate w_{+m} minus the detachment rate w_{-m} multiplied by the probability to find the unit m at the end of the chain:

$$v\mu(m) = w_{+m} -w_{-m}\mu(m).$$

Inverting this relationship, the probability is obtained as:

$$\mu(m) = \frac{w_{+m}}{w_{-m} + v}.$$

Because this probability distribution should be normalized to unity, we find the following self-consistent equation for the growth velocity in terms of the attachment and detachment rates:

$$\sum_{m=1}^{M}\mu(m) = \sum_{m=1}^{M}\frac{w_{+m}}{w_{-m}+v} = 1.$$

The figure below illustrates the copolymerization of two monomeric species if the kinetics obeys the law of mass action, according to which the attachment rates are proportional to the concentration of the attached monomer, $w_{+m}=k_{+m}c_{m}$; and the detachment rates are independent of the concentrations, $w_{-m}=k_{-m}$. In the concentration space, equilibrium happens on the line $\sum_{m=1}^{M}(k_{+m}/k_{-m})c_{m} = 1$, as shown in the figure below. The copolymer is growing for larger concentrations and it undergoes depolymerization for smaller concentrations. In between, the growth velocity is vanishing, but the copolymer length fluctuates because of random attachment and detachment events.

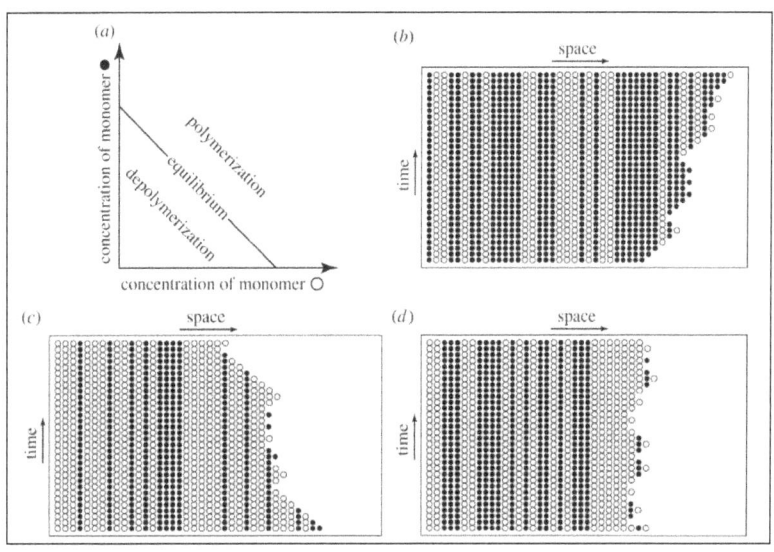

The above figure shows Growth of Bernoulli chains for the rate constants $k_{\pm 1} = k_{\pm 2} = 1$ and the concentration $c_2 = 0.5$. The monomers 1 are depicted by filled circles and the monomers 2 as open circles. (a) The space of monomeric concentrations where equilibrium happens on the line $c_1 + c_2 = 1$. (b) Space–time plot of copolymerization if $c_1 = 1$. (c) Space–time plot of depolymerization if $c_1 = 0.25$. (d) Space–time plot at equilibrium if $c_1 = 0.5$.

The figure below shows another example with rates depending on an external force f as:

$$w_{+m} = k^0_{+m} e^{\beta f \delta_{+m}} c_m, \ w_{-m} = k^0_{-m} e^{\beta f \delta_{-m}}, \ (m = 1, 2)$$

$$\text{where } \beta = (k_B T)^{-1}, \delta_{\pm 1} = \delta_{\pm 2} = \pm \delta / 2,$$

$$k^0_{+1} = k^0_{+2} = 1, k^0_{-1} = 0.01, \ k^0_{-2} = 0.001, \ c_1 = 0.01 \text{ and } c_2 = 0.0005.$$

As seen in the figure above, the growth is stopped if a strong enough force is opposed, which defines the stall force: $\beta f_{st} \delta = \ln\left(\frac{2}{3}\right) \simeq -0.4055$. The growth is driven by sequence disorder if $-0.4055 < \beta f \delta \leq 0.09878$ and by a positive free energy if $0.09878 < \beta f \delta$. The figure above shows that the copolymer can grow against an opposed external force, while the corresponding pure polymers cannot because their velocity v becomes positive only at positive values of the rescaled external force $\beta f \delta$. Accordingly, a mechanical work is generated by the sequence disorder of the growing copolymer, but not with the pure polymers.

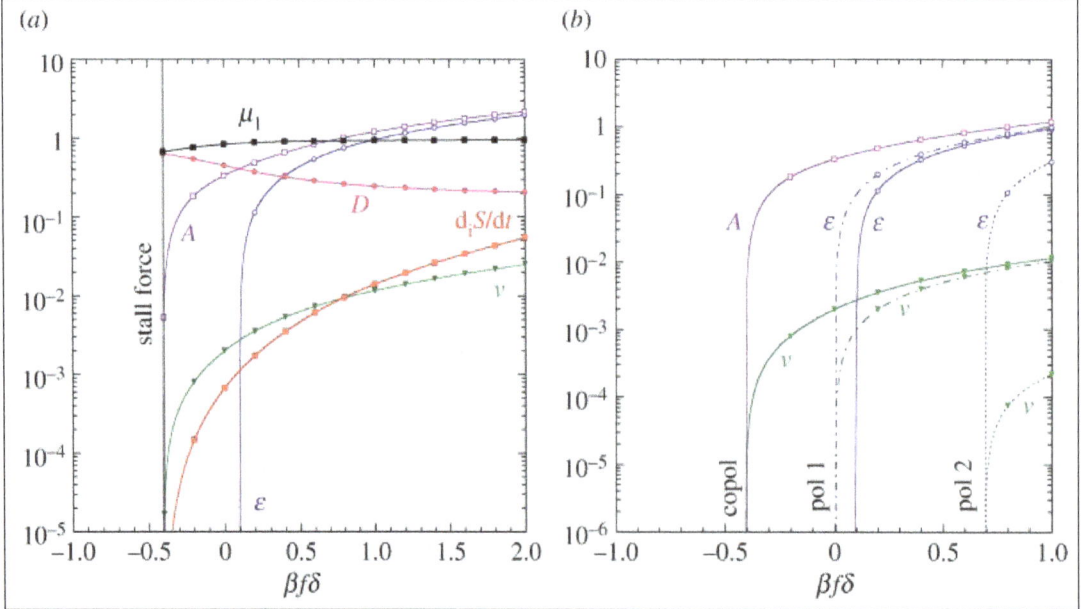

The above figure shows: (a) Growth of Bernoulli chains in the conditions (2.13) versus the rescaled external force $\beta f \delta$. v is the growth velocity (filled triangles), μ_1 the fraction of monomeric units m=1 in the sequence (filled squares), D the Shannon disorder (filled circles), ε the free-energy driving power (open circles), A=ε+D the affinity (open squares) and $d_i S/dt = vA$ the entropy production (crossed squares). (b) Growth of Bernoulli chains in the conditions (solid lines and symbols) and pure polymers with different concentrations: for the polymer 1, the concentrations are $c_1 = 0.01$ and $c_2 = 0$ (dashed-dotted lines and symbols). For the polymer 2, the concentrations are $c_1 = 0$ and

$c_2 = 0.0005$ (dashed lines and symbols). The growth velocity v (filled triangles), the free-energy driving power ε (open circles) and the affinity $A = \varepsilon + D$ (open squares) are plotted versus the rescaled external force $\beta f\delta$. The symbols are the results of kinetic Monte Carlo simulations and the solid lines show the theoretical predictions of equations $\mu(m) = \dfrac{w + m}{w_{-m} + v}$ and $\displaystyle\sum_{m=1}^{M} \mu(m) = \sum_{m=1}^{M} \dfrac{w_{+m}}{w_{-m} + v} = 1.$

Copolymer Composition

The successful synthesis of new materials via free radical copolymerization requires a thorough understanding of the factors that control the structures of copolymer chains. Of the primary structural variables used to describe polymer chains, two (copolymer composition and comonomer sequence distribution) are unique to copolymers. The compositions and sequences of copolymers prepared via free-radical processes.

The products of free radical copolymerizations are, with few exceptions, determined by the kinetics, rather than the thermodynamics, of the chain growth process. The problem of predicting copolymer composition and sequence then reduces to the writing of a set of differential equations that describe the rates at which each of the two monomers enters the copolymer chain by attack of the growing macroradical. This requires a kinetic model of the copolymerization process, and several such models have been described in the copolymerization literature. The following sections examine the fundamental bases of the most important of these models, and assess the degree to which such models can account for experimentally observed copolymerization behavior.

Terminal Model

The standard kinetic treatment of free radical copolymerization was introduced in 1944, in papers contibuted independently by Mayo and Lewis, by Alfrey and Goldfinger and by Wall Following earlier suggestions by Dostal, by Norrish and Brookman, and by Jenckel, Mayo and Lewis described their experimental work on the radical copolymerization of styrene and methyl methacrylate in terms of a model in which the rate constant for addition of each monomer was assumed to be dependent on the identity of the terminal unit on the growing chain. Four elementary propagation steps were then considered in equation below:

$$-M_1\cdot \quad + \quad M_1 \xrightarrow{\ k_{11}\ } -M_1M_1\cdot$$

$$-M_1\cdot \quad + \quad M_2 \xrightarrow{\ k_{12}\ } -M_1M_2\cdot$$

$$-M_2\cdot \quad + \quad M_1 \xrightarrow{\ k_{21}\ } -M_2M_1\cdot$$

$$-M_2\cdot \quad + \quad M_2 \xrightarrow{\ k_{22}\ } -M_2M_2\cdot$$

By writing differential equations that describe the rates of disappearance of monomers M1 and M2, and by assuming steady-state concentrations of the radical centers M_{1^*} and M_{2^*}, one arrives at a simple expression that relates the ratio of monomers in the copolymer $(d[M_1]/d.[M_2])$ to the concentrations of monomers in the feed mixture ($[M_1]$ and $[M_2]$):

$$\frac{d[M_1]}{d[M_2]} = \frac{[M_1]}{[M_2]} \frac{r_1[M_1] + [M_2]}{[M_1] + r_2[M_2]}$$

The parameters r_1 and r_2 are reactivity ratios defined as:

$$r_1 = k_{11}/k_{12} \quad \text{and} \quad r_2 = k_{22}/k_{21}$$

Analysis of the terminal model is readily extended to the prediction of copolymer sequence distribution. Sequence distributions are most generally and conveniently specified in terms of the number fractions of uninterrupted sequences of a given monomer (M_1 or M_2) that are of a particular length. The number fraction is of course identical to the probability that a given uninterrupted sequence, selected at random, is of that length. Consider a sequence of M1 units of length x. Such a sequence arises in the terminal model when a growing macroradical terminating in M_{1*} adds (x - 1) M_1's followed by an M2. The probability that such a sequence forms is obtained as the product of the probabilities of each of the independent steps that lead to the sequence. Thus the number fraction of sequences of M_1 of length x (Nx^1) is given as:

$$N_x^1 = P_{11}^{(x-1)} P_{12}$$

where P_{11} is the probability that $-M_{1*}$ adds M_1:

$$P_{11} = \frac{k_{11}[M_1 \cdot][M_1]}{k_{11}[M_1 \cdot][M_1] + k_{12}[M_1 \cdot][M_2]} = \frac{r_1}{r_1 + [M_2]/[M_1]}$$

and P_{12} ($= 1 - P_{11}$) is the probability that $-M_1^*$ adds M2. The lengths of M2 sequences are determined in similar fashion, such that:

$$N_x^2 = P_{22}^{(x-1)} P_{21} = \frac{r_2}{r_2 + [M_1]/[M_2]}$$

Thus Equations $P_{11} = \dfrac{k_{11}[M_1 \cdot][M_1]}{k_{11}[M_1 \cdot][M_1] + k_{12}[M_1 \cdot][M_2]} = \dfrac{r_1}{r_1 + [M_2]/[M_1]}$ and $N_x^2 = P_{22}^{(x-1)} P_{21} = \dfrac{r_2}{r_2 + [M_1]/[M_2]}$

allow calculation of the comonomer sequence distribution from a knowledge of the terminal model

reactivity ratios and the monomer feed composition. Use of Equations $\dfrac{d[M_1]}{d[M_2]} = \dfrac{[M_1]}{[M_2]} \dfrac{r_1[M_1] + [M_2]}{[M_1] + r_2[M_2]}$,

$P_{11} = \dfrac{k_{11}[M_1 \cdot][M_1]}{k_{11}[M_1 \cdot][M_1] + k_{12}[M_1 \cdot][M_2]} = \dfrac{r_1}{r_1 + [M_2]/[M_1]}$ and $N_x^2 = P_{22}^{(x-1)} P_{21} = \dfrac{r_2}{r_2 + [M_1]/[M_2]}$ is of course re-

stricted to conditions under which the monomer feed composition is fixed. In practical terms, this requires the use of low monomer conversions in order to avoid serious errors arising from compositional drift as M1 and M2 enter the copolymer chain at different rates. Use of the integrat-

ed form of Equation $\dfrac{d[M_1]}{d[M_2]} = \dfrac{[M_1]}{[M_2]} \dfrac{r_1[M_1] + [M_2]}{[M_1] + r_2[M_2]}$ has been recommended in order to relax the

low-conversion restriction.

Penultimate Model

The fundamental assumption of the terminal model, i.e., that the reactivity of the growing radical is determined only by the identity of the last-added monomer unit, is equivalent to the assumption that the relative rates of monomer addition are insensitive to substitution at positions more remote than that β to the radical center. Remote substituent effects are well known in organic chemistry, and it is plausible that copolymerization reactivity ratios should be affected by units that precede the terminal residue on the propagating macroradical. Merz, Alfrey and Goldfinger suggested in 1946 that a proper description of the propagation step should take into account four distinct active centers, which are defined by the identities of their teminal and penultimate units:

$$-M_1M_1\cdot \ + \ M_1 \xrightarrow{\ 111\ } -M_1M_1M_1\cdot$$

$$-M_1M_1\cdot \ + \ M_2 \xrightarrow{\ k_{112}\ } -M_1M_1M_2\cdot$$

$$-M_2M_1\cdot \ + \ M_1 \xrightarrow{\ k_{211}\ } -M_2M_1M_1\cdot$$

$$-M_1M_2\cdot \ + \ M_2 \xrightarrow{\ k_{212}\ } -M_2M_1M_2\cdot$$

$$-M_1M_2\cdot \ + \ M_1 \xrightarrow{\ k_{121}\ } -M_1M_2M_1\cdot$$

$$-M_1M_2\cdot \ + \ M_2 \xrightarrow{\ k_{122}\ } -M_1M_2M_2\cdot$$

$$-M_2M_2\cdot \ + \ M_1 \xrightarrow{\ k_{221}\ } -M_2M_2M_1\cdot$$

$$-M_2M_2\cdot \ + \ M_2 \xrightarrow{\ k_{222}\ } -M_2M_2M_2\cdot$$

The copolymer composition is determined by the relative rates of monomer consumption:

$$\frac{d[M_1]}{d[M_2]} = \frac{k_{111}[M_1M_1\cdot][M_1] + k_{211}[M_2M_1\cdot][M_1]+k_{121}[M_1M_2\cdot][M_1]+K_{221}[M_2M_2\cdot][M_1]}{k_{112}[M_1M_1\cdot][M_2] + k_{212}[M_2M_1\cdot][M_2] + k_{122}[M_1M_2\cdot][M_2]+k_{222}[M_2M_2\cdot][M_2]}$$

Assumption of steady-state concentrations of each of the four radical centers leads to Equation below for the copolymer composition:

$$\frac{d[M_1]}{d[M_2]} = \frac{1+\dfrac{r_{21}X(r_{11}X+1)}{r_{21}X+1}}{1+\dfrac{r_{12}(r_{22}+X)}{X(r_{12}+X)}}$$

where X = $[M_1]/[M_2]$ and the reactivity ratios are defined as:

$$r_{11} = \frac{k_{111}}{k_{112}} \quad r_{21} = \frac{k_{211}}{k_{212}} \quad r_{12} = \frac{k_{122}}{k_{121}} \quad r_{22} = \frac{k_{222}}{k_{221}}.$$

Prediction of monomer sequence lengths by the penultimate model is conceptually identical to that described previously for the terminal model. The probability (P_{211}) that an $-M_2M_1*$ chain end adds M_1 is:

$$P_{211} = \frac{k_{211}[M_2M_1\cdot][M_1]}{k_{211}[M_2M_1\cdot]Mil + k_{212}[M_2M_1\cdot][M_2]} - \frac{[M_1]}{[M_1]+[M_2]/r_{21}}$$

and

$$P_{111} = \frac{[M_1]}{[M_1]+[M_2]/r_{11}}$$

A sequence consisting of an isolated M1 unit arises only when an -M2M$_1$* chain end adds M$_2$, so the number fraction of M_1 sequences that are of length 1 is:

$$N_1^1 = 1 - P_{211}$$

For longer sequences, enumeration of the required propagation steps leads to:

$$N_x^1 = P_{211}P_{111}^{(x-2)}(1-P_{111})$$

Eqns $\dfrac{d[M_1]}{d[M_2]} = \dfrac{1+\dfrac{r_{21}X(r_{11}X+1)}{r_{21}X+1}}{1+\dfrac{r_{12}(r_{22}+X)}{X(r_{12}+X)}}$, $\quad N_1^1 = 1 - P_{211}$ and $\quad N_x^1 = P_{211}P_{111}^{(x-2)}(1-P_{111})$ for the pen-

ultimate model are thus equivalent to Equations $\dfrac{d[M_1]}{d[M_2]} = \dfrac{[M_1]}{[M_2]}\dfrac{r_1[M_1]+[M_2]}{[M_1]+r_2[M_2]}$ and

$P_{11} = \dfrac{k_{11}[M_1\cdot][M_1]}{k_{11}[M_1\cdot][M_1]+k_{12}[M_1\cdot][M_2]} = \dfrac{r_1}{r_1+[M_2]/[M_1]}$ developed previously for the terminal

model; in each case, knowledge of the copolymerization reactivity ratios allows calculation of co-polymer compositions and sequences as functions of the ratio of monomer concentrations in the feed.

Complex Participation Model

Radical copolymerizations of electron-rich olefins with electron-poor olefms are anomalous in several respects. Such monomer pairs often afford alternating copolymers over the entire range of feed cornposition and one often observes in such systems a marked sensitivity of the overall copolymerization rate to temperature, solvent and monomer concentration. Butler and coworkers have also noted anomalies in the stereochemistry and region chemistry of certain copolymerizations of electron-rich and electron-poor olefins.

A mechanistic scheme that accounts for this behavior invokes the participation of 1: 1 olefinic electron donor-acceptor (EDA) complexes in the propagation step. Specifically, it is proposed that - the 1: 1 complex $\left(\overline{M_1M_2}\right)$ competes with free monomers for the growing chain end. Modification of the terminal model in this way requires consideration of eight propagation steps:

$$-M_1\cdot + \quad M_1 \xrightarrow{k_{11}} \quad -M_1M_1\cdot$$

$$-M_1 \cdot + \quad M_2 \xrightarrow{k_{12}} \quad -M_1M_2 \cdot$$

$$-M_2 \cdot + \quad M_1 \xrightarrow{k_{21}} \quad -M_1M_2 \cdot$$

$$-M_2 \cdot + \quad M_2 \xrightarrow{k_{22}} \quad -M_2M_2 \cdot$$

$$-M_1 \cdot + \quad \overline{M_1M_2} \xrightarrow{k_{112}} -M_1M_1M_2 \cdot$$

$$-M_1 \cdot + \quad \overline{M_2M_1} \xrightarrow{k_{121}} -M_1M_2M_1 \cdot$$

$$-M_2 \cdot + \quad \overline{M_1M_2} \xrightarrow{k_{212}} -M_2M_1M_2 \cdot$$

$$-M_2 \cdot + \quad \overline{M_2M_1} \xrightarrow{k_{221}} -M_2M_2M_1 \cdot$$

If radical additions to each "side" of the complex are regarded as distinct, and a complexation equilibrium:

$$M_1 + M_2 \xrightleftharpoons{K} \overline{M_1M_2}$$

An analysis of this model, which predicts copolymer composition and sequence as functions of the feed composition, has been provided by Hill and coworkers. The mole ratio of M_1 to M_2 in the copolymer is given as:

$$\frac{d[M_1]}{d[M_2]} = \frac{(1-P_{22})(P_{12}+P_{1\overline{12}})+(1-P_{12})(P_{21}+P_{2\overline{21}})}{(1-P_{21})(P_{12}+P_{1\overline{12}})+(1-P_{11})(P_{21}+P_{2\overline{21}})}$$

Where the transition probabilities are defined as:

$$p22 = r_2[M_2]/\sum M_2 \qquad\qquad P_{21} = [M_1]/\sum M_2$$

$$P_{212} = s_2[\overline{M_1M_2}]/\sum M_2 \qquad P_{221} = s_2q_2[\overline{M_1M_2}]/\sum M_2$$

$$P_{12} = [M_1]/\sum M_1 \qquad\qquad P_{11} = r_1[M_1]/\sum M_1$$

$$P_{112} = s_1q_1[\overline{M_1M_2}]/\sum M_1 \qquad P_{121} = s_1[\overline{M_1M_2}]/\sum M_1$$

With

$$\sum M_1 = [M_1]+r_1[M_1] + s_1[\overline{M_1M_2}][1+q_1]$$

and

$$\sum M_2 = r_2[M_2] + [M_1] + s_2[\overline{M_1M_2}][1+q_2]$$

The reactivity ratios in this formulation are defined as:

$$r_1 = k_{11}/k_{12} \qquad\qquad r_2 = k_{22}/k_{21}$$

$$q_1 = k_{1\overline{12}}/k_{1\overline{21}} \qquad\qquad 92 = k_{2\overline{21}}/k_{2\overline{12}}$$

$$s_1 = k_{1\overline{21}}/k_{12} \qquad\qquad s_2 = k_{2\overline{12}}/k_{21}$$

so that Equation $\dfrac{d[M_1]}{d[M_2]} = \dfrac{(1-P_{22})(P_{12}+P_{1\overline{12}})+(1-P_{12})(P_{21}+P_{2\overline{21}})}{(1-P_{21})(P_{12}+P_{1\overline{12}})+(1-P_{11})(P_{21}+P_{2\overline{21}})}$ specifies the copolymer com-

position in terms of monomer concentrations and seven parameters (six reactivity ratios and the complexation equilibrium constant, K).

Sequence information can be calculated in the usual manner, i.e., as the number fraction of sequences of either monomer of length x. For M_1, the number fraction of sequences of length x is:

$$N_1{}^x = \frac{p_1{}^x}{\displaystyle\sum_{m=1}^{\infty} p_1{}^m}$$

where

$$p_1{}^1 = p_2 \frac{[(1-P_{11})(P_{12}+P_{1\overline{21}})(P_{21}+P_{2\overline{21}})+P_{2\overline{12}}(P_{1\overline{12}}(P_{1\overline{12}}+P_{12})]}{(P_{1\overline{12}}+P_{12})}$$

and

$$p_1{}^x = P_2 P_{11}{}^{(x-2)} \frac{[(1-P_{11})(P_{21}+P_{2\overline{11}})(P_{12}P_{11}+P_{1\overline{21}}P_{11}+P_{1\overline{12}})}{(P_{1\overline{12}}+P_{12})}$$

The quantity P_2 is the probability of selecting an M_2 unit that entered the chain either as free M_2 or via the reactions shown in Equations $-M_1\cdot + \overline{M_1 M_2} \xrightarrow{k_{112}} -M_1M_1M_2$ and $-M_2\cdot + \overline{M_1 M_2} \xrightarrow{k_{212}} -M_2M_1M_2\cdot$ However, P_2 need not be evaluated, since this quantity may be eliminated from the expression for N_x1. Eqns $\dfrac{d[M_1]}{d[M_2]} = \dfrac{(1-P_{22})(P_{12}+P_{1\overline{12}})+(1-P_{12})(P_{21}+P_{2\overline{21}})}{(1-P_{21})(P_{12}+P_{1\overline{12}})+(1-P_{11})(P_{21}+P_{2\overline{21}})}$

and $N_1{}^x = \dfrac{p_1{}^x}{\displaystyle\sum_{m=1}^{\infty} p_1{}^m}$ thus allow calculation of copolymer composition and sequence as functions of the monomer feed composition.

Other Copolymerization Models

The terminal, penultimate and complex participation models have been discussed widely in the copolymerization literature. Two additional models - the complex dissociation model and the depropagation model - have not been considered as extensively, but each is physically plausible and each has been analyzed in sufficient detail that compositions and sequences may be calculated.

These models are outlined briefly here; the reader is directed to the original papers for a thorough description.

Complex Dissociation Model - Tsuchida and Tomono suggested in 1971 that EDA complexes may take part in radical copolymerizations not by adding to the chain end in a concerted fashion, but rather by delivering only one of the two complexed monomers. Thus the terminal model must be modified by consideration of four new propagation steps:

$$-M_1\cdot \ + \ \overline{M_1M_2} \rightarrow M_1M_1\cdot \ + \ M_2$$

$$-M_1\cdot \ + \ \overline{M_1M_2} \rightarrow M_1M_2\cdot \ + \ M_1$$

$$-M_2\cdot \ + \ \overline{M_1M_2} \rightarrow M_2M_1\cdot \ + \ M_2$$

$$-M_2\cdot \ + \ \overline{M_1M_2} \rightarrow M_2M_2\cdot \ + \ M_1$$

Hill and coworkers have provided an analysis of this kinetic scheme, and have demonstrated the calculation of copolymer composition and sequence according to this model.

Copolymerization with Depropagation - Most radical copolymerizations are strongly exothermic and effectively irreversible. Near the ceiling temperature, however, the influence of depropagation must be considered. Lowry in 1960 developed a general theory that predicts copolymer composition for systems in which the addition of one of the two monomers is reversible. O'Driscoll and coworkers subsequently derived composition equations equivalent to but more general than those of Lowry, and provided expressions for sequence distributions as well.

Evaluation of Copolymerization Models

Three experimental approaches have been used to evaluate the theoretical treatments of copolymerization discussed above. The majority of such studies have compared measured compositions and sequences with those predicted by each of the kinetic models. But in fact, the determination of sequence distributions is still a considerable technical challenge, so that many investigators have been limited to composition measurements alone. More recently, measurements of absolute rate constants and trapping on simple model radicals have been brought to bear on questions of copolymerization mechanism.

Composition and Sequence. It was recognized early in the study of radical copolymerization that sequence distribution should be more sensitive than composition to the details of the chain growth process. In their original paper on the penultimate model, Merz, Alfrey and Goldfinger pointed out that it is not in measurements of composition, but rather "in the length of the comonomer sequences that the effect of the monomer in the chain preceding the fiee-radical chain end would become noticeable" Berger and kuntzl in 1964 analyzed this problem quantitatively for several hypothetical and several real copolymerizations. They showed, for example, that the compositions predicted by the terminal and penultimate models would be indistinguishable over a 1000-fold variation in $[M_1]/[M_2]$, for a system in which the terminal model reactivity ratios are $r_1 = 0.1$ and $r_2 = 0.9$ and the penultimate model parameters are $r_{11} = 0.94, r_{21} = 0.01$, $r_{22} = 0.9$ and 1-12 = 5. Thus quite large penultimate effects (r_1/r_{21} = 94) can be masked. On the other hand, much more modest effects are readily apparent in the predicted sequence distributions. Analysis of the copolymerization of styrene (M_1) and maleic anhydride (M_2) according to the terminal (r_1 = 0.0227, r_2 = 0) or penultimate (r_{11} = 0.0 17, r_{21} = 0.063, r_{12} = r_{22} = 0) models leads to rather different predictions. In the terminal model analysis, the number fraction of styrene residues isolated between maleic anhydride units is 0.23; in the penultimate model analysis, 0.09. Thus a penultimate effect of a factor of five leads to sequence predictions markedly different from those of the terminal model.

More recently, Hill and coworkers have analyzed the bulk copolymerization of styrene and acrylonitrile in terms of the composition and sequence predictions of the terminal, penultimate and complex participation models. They find that all three models reproduce the experimental composition data rather well, although the penultimate and complex models offer statistically significant improvements over the terminal model. Figure shows the experimental compositions as well as the best-fit predictions of each of the three models. The predictions are remarkably similar, and data of very high precision are required for model discrimination. The sequence predictions of the three models are quite different, however, and allow a clear distinction between the penultimate and complex kinetic schemes. Hill and coworkers conclude that the bulk copolymerization of styrene and acrylonitrile is best described by a penultimate model with rss = 0.23, rAS = 0.63, rSA = 0.09 and rAA = 0.04.

Determination of copolymer sequence has also provided insight into the effects of solvent on radical copolymerization. Hanvood and coworkers have noted that copolymerizations that involve ionic, highly polar or hydrogen-bonding monomers are subject to large solvent effects; (i.e., the composition curves for such copolymerizations vary dramatically with solvent). Furthermore, sequence distributions determined in such systems are inconsistent with the predictions of any of the conventional kinetic schemes, if one uses reactivity ratios determined from the relation between copolymer composition and monomer feed composition. On the other hand, comparison of the sequence distributions of copolymers of identical composition (but prepared in different solvents from feeds of different $[M_1]/[M_2]$) shows them to be identical. Harwood concludes that the conditional probabilities governing monomer addition (and therefore the reactivity ratios) must be independent of solvent and that the role of the solvent is to influence the relative concentrations of monomers available to the growing chain end. Hanvood has provided convincing evidence for this behavior in the copolymerizations of styrene with acrylic acid, methacrylic acid and acrylamide, and in the copolymerization of vinylidene chloride with methacrylonitrile.

Measurements of Absolute Rate Constants

Each of the copolymerization models predicts not only composition and sequence, but also the overall propagation rate constant, k_p, as a function of monomer feed composition. Fukuda and coworkers used a rotating sector technique to determine k_p over a range of feed compositions for the copolymerization of styrene and methyl methacrylate. They found large and systematic deviations from the predictions of the terminal model, but were able to reproduce their experimental observations by assuming a small penultimate effect. This is an intriguing observation in view of the "classic" nature of the styrene/methyl methacrylate copolymerization, and suggests that absolute rate measurements may prove particularly powerful in probing copolymerization mechanism.

Model Reactions

The mechanistic assumptions of the terminal, penultimate and complex participation models may be evaluated via trapping with simple alkyl radicals. For example, the penultimate model, as applied to the copolymerization of monosubstituted olefins, implies that the selectivity of the attacking radical should be sensitive to the nature of the substituent that lies γ to the radical center:

$$-CH_2CHCH_2CH\cdot$$
$$\quad\quad |\quad\quad\quad |$$
$$\quad\quad \tilde{a}\quad\quad\quad \alpha$$

Tirrell and coworkers have determined the relative rates of addition of acrylonitrile and styrene (k_A/kS) to a series of y-substituted propyl radicals and report that a y-cyano group depresses the relative affinity of the radical center for acrylonitrile by a factor of 3.5. This "penultimate effect" is remarkably consistent with those inferred by Hill and coworkers via composition and sequence analyses and lends support to the penultimate model as a physically realistic description of the copolymerization of styrene and acrylonitrile. Analogous measurements of k_A/kS for the l-phenylethyl and 1-cyanoethyl radicals are consistent with this view.

Trapping have also been applied to the evaluation of the complex participation model. The hypothesis that olefinic EDA complexes add in concerted fashion to alkyl radicals - the fundamental assertion of the model - is subject to direct experimental test, as shown in Equations below.

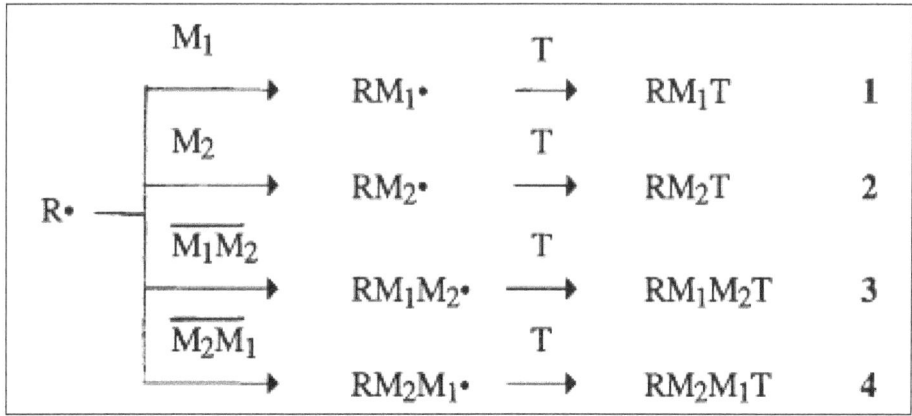

T = radical trap

The radical of interest, R•, is generated in the presence of M_1, M_2 and a radical trap (T). If R• undergoes concerted complex addition, trapping of the simple olefin adducts RM_1• and RM_2• will not be observed. Determination of the yields of products 1 and 2 then allows an estimate of the maximum extent to which the complex participates in the consumption of M_1 and M_2. Jones and Tirrell have reported trapping on the 1-butyl radical in its reactions with Nphenylmaleirnide and two donor olefins (2-chloroethyl vinyl ether and styrene). In each case, simple addition of N-phenylmaleimide was the dominant reaction; no evidence for concerted addition of the EDA complex was obtained.

Table: Monomer Reactivity Ratios in Radical Copolymerizations.

M$_1$	r$_1$	M$_2$	r$_2$
Acrylonitrile	0.030-0.100	Butadiene	0.10-0.45
	7.00	Ethylene	0.00
	6.00	Maleic Anhydride	0.00
	0.14	Methyl Methacrylate	1.32
	0.00-0.17	Styrene	0.29-0.55
	4.05-5.51	Vinyl Acetate	0.040-0.060
	2.55-4.00	Vinyl Chloride	0.020-0.070
Butadiene	0.50-0.75	Methyl Methacrylate	0.027-0.32
	1.35-1.83	Styrene	0.37-0.84
	8.80	Vinyl Chloride	0.04
Ethylene	0.40	Maleic Anhydride	0.00
	0.050	Styrene	14.9
	0.13-0.88	Vinyl Acetate	0.72-3.74
	0.020-0.34	Vinyl Chloride	0.96-4.38
Maleic Anhydride	0.010-0.020	Methyl Methacrylate	3.10-6.36
	0.000-0.020	Styrene	0.000-0.097
	0.40-067	Vinyl Chloride	0.040-0.100
Styrene	18.8-60.0	Vinyl Acetate	0.010-0.16
		Vinyl Chloride	0.005-0.160
Vinyl Acetate	0.24-0.98	Vinyl Chloride	1.03-2.30

Reactivity Ratio

Since the development of Mayo-Lewis equation for binary copolymerisation based on the terminal kinetic model, a large volume of literature has been devoted to it. This large quantity of experimental data has mainly been used to determine reactivity ratio, resonance stabilisation (Q) and electronegativity (e) parameters. The kinetics and mechanisms involved in the free-radical binary copolymerisation of vinyl monomers have been a research topic of interest for many years. The composition and sequence distribution of a copolymer chain are dependent on the relative proportion of applied monomers as well as the monomer and radical reactivities.19 Different models have been put forth to visualise the mechanism of addition of growing chains and the factors influencing them.

Estimation of reactivity ratios is important because of the following reasons. Primarily to predict copolymer composition and microstructure for any starting mixture, secondly to classify the relative reactivities of different monomers toward free macro-radicals and finally to understand issues related to the rate of copolymerisation and molecular weight distribution.

Copolymerisation reactivity ratios were originally measured for the purpose of describing the relative reactivities of various monomers towards various radicals. Now-a-days it is treated as quantitative data and hence the necessity of accurate measurement arises. Several methods have been developed to estimate reactivity ratios. Some of these methods require labourious calculation procedures and so computer is put to use in an efficient manner to estimate reactivity ratios.

The estimation methods developed to determine reactivity ratio are based on the binary copolymer composition equations $\dfrac{m_1}{m_2} = r_2 = \dfrac{r_1[M_1]^2 + [M_1][M_2]}{r_2[M_2]^2 + [M_1][M_2]}$ and equation below:

$$\frac{dM_1}{dM_2} = \frac{r_1 M_1^2 + M_1 M_2}{r_2 M_2^2 + M_1 M_2},$$

where r_1 and r_2 are the monomer reactivity ratios and dM_1/dM_2 is the relative rate of addition of the two monomers to the chain. The copolymer composition may not be independent of conversion. This means the disappearance of monomer one may be faster than the disappearance of monomer two, if monomer one is being incorporated into the copolymer at a faster rate and therefore has a larger reactivity ratio than monomer two.

Depending on the reactivity ratios of the monomers, the copolymer can incorporate the comonomers in different ways. The three main types of behaviour that copolymerisations tend to follow correspond to the conditions when both r_1 and r_2 are equal to one, $r_1 \cdot r_2 < 1$ and $r_1 \cdot r_2 > 1$.

A perfectly random copolymerisation is achieved when both r_1 and r_2 values are equal to one. This type of copolymerisation will occur when the two different types of propagating species, $-M_1 \cdot$ and $-M_2 \cdot$ show the exact same preference for addition of each type of monomer.

An alternating copolymerisation is defined as $r_1 = r_2 = 0$. The polymer product in this type of copolymerisation shows a non-random equimolar amount of each comonomer that is incorporated into the copolymer. This may occur because the growing radical chains will not add to its own monomer. Therefore, the opposite monomer will have to be added to produce a growing chain and a perfectly alternating chain is obtained.

When $r_1 > 1$ and $r_2 > 1$, both of the monomers will add to themselves and in theory could produce block copolymers. But due to short lifetime of propagating radical, such copolymerisation produces undesirable heterogeneous products that include homopolymers. Therefore, macroscopic phase separation could occur and desirable physical properties such as transparency would not be obtained.

In order to determine the amount of comonomer that has been incorporated into the copolymer, various analytical methods must be used. Proton Nuclear Magnetic Resonance, Carbon 13 NMR, Chromatographic techniques such as Gas chromatography / HPLC and Fourier Transform Infrared Spectroscopy etc. are the sensitive instrumental techniques which are helpful in determining copolymer composition.

The copolymer composition equation is mostly employed to determine monomer reactivity ratios for low conversion copolymerisations. The copolymerisation systems are characterised by their monomer reactivity ratios under particular set of conditions (solvent, temperature etc.). Different methods, weighing in accuracy and ease of operation, have come into existence for their determination.

Methods to estimate reactivity ratios are classified as differential of integral depending on whether they started from the differential or integral form of the MayoLewis equation. Many methods have been used to estimate reactivity ratios of a large number of comonomers. Tidwell and Mortimer distinguished four different procedures for the calculation of monomer reactivity ratios in copolymerisation, via. linearisation, approximation, intersection, and curve fitting. We will quickly review few of them which still have importance in today's context.

Linear Methods

Intersection Method

So as to determine the reactivity ratio of binary copolymerisations, a mathematical solution can be reached by performing two experiments. Equation $\dfrac{d[M_1]}{d[M_2]} = r_2 = \dfrac{r_1[M_1]^2 + [M_1][M_1]^2}{r_2[M_2]^2 + [M_1][M_1]^2}$ is expressed into the following linear equation given below:

$$r_2 = \frac{[M_1]}{[M_2]}\left[\frac{m_1}{m_2} - 1\right] + \frac{m_2[M_1]^2}{m_1[M_2]^2} r_1$$

Where $[M_1]$, $[M_2]$ are the molar concentrations in the feed and m_1 and m_2 in the copolymer. A graphic representation of above equation gives a set of straight lines in r_1, r_2 coordinate, each line representing one experiment. The point whose coordinates are located at the intersection of two different experiments represents r_1 and r_2 values.

Joshi-Kapur Method

As seen that points resulting from the intersection of lines at very small angles are further away from the hypothetically correct value, a weighting of the intersection points has been attempted using a function of the intersection angles where θ is the angle of intersection of the two lines.

$$r_i = \frac{\sum r_i[tg\theta]}{\sum [tg\theta]}$$

$$r_i = \frac{\sum r_i[\sin\theta]}{\sum [\sin\theta]}$$

The Joshi-Kapur method considers all $n(n-1)/2$ intersections of pair combinations of all 'n' experimental lines. So by this way the intersection points are also evaluated. These average values of r_i, estimated according to empirical criteria, do not assume that one of the points as best but expresses a possibility that another point may be a better fit to the experimental data than any of the intersection points.

Joshi and Joshi Method

Another method was put forwarded by Joshi and Joshi which eliminates the subjective element in selection of the best point of intersection by statistically finding the closest point to all experimental lines. This method is an improvement of the Mayo Lewis intersection method resulting ideally in an unique intersection point. A condition was set up to make the sum of the squares of the perpendicular distances from all the lines to the best point of intersection to a minimum. This method does not suffer from the errors of reindexing the monomers.

Fineman-Ross Method

The first method to balance successfully the entire set of experiments and to estimate the experimental error was developed by Fineman and Ross in 1950. They reformulated the Mayo-Lewis equation so that the data points result in a single straight line. This method offers a simple graphical evaluation of reactivity ratios. A simpler method of estimation would involve carrying out the copolymerisations to low conversions and using the approximate form of differential copolymerisation equation to estimate the reactivity ratios. However, a simpler technique, which permits the use of data in the intermediate concentration regions and reduces the uncertainties in the r values,

is possible. If f = (m_1/m_2) and F = (M_1/M_2), then the differential equation $\dfrac{dM_1}{dM_2} = \dfrac{r_1 M_1^2 + M_1 M_2}{r_2 M_2^2 + M_1 M_2}$ can be rewritten as:

$$\frac{F(1-f)}{f} = r_2 - \frac{F^2}{f} r^1$$

Equation above can be represented as:

$$G = r_1 H - r_2$$

where G = (f -1)/F and H = f/F² . A plot of G as ordinate and H as abscissa is a straight line whose slope is - r_1 and intercept is r_2. The method of least squares can be employed to find the line of best fit. The slope of the line of best fit is influenced to a great extent by the points which are nearer to the origin, thus giving a non-uniform weightage to the points and hence suffering from errors of reindexing the monomers. The validity is only qualitative and the estimates of r_1 and r_2 can change with each experiment by weighing the data in different ways. Furthermore, the high and low experimental composition data are unequally weighed, which produces large effects on the calculated values of r_1 and r_2.

Yezrielev-Brokhina-Roskin Method

As in the intersection method there exist a lack of symmetry, Fineman-Ross equation leads to different values of reactivity ratios even though the same equation and the same experimental data are used. To overcome this lack of symmetry Yezrielev et al. derived the symmetric equation below:

$$\left[1 - \frac{m_2}{m_1}\right]\sqrt{\frac{m_1}{m_2}} = \frac{M_1}{M_2} = \sqrt{\frac{m_1}{m_2}}r_1 - \frac{M_1}{M_2}\sqrt{\frac{m_1}{m_2}}r_2$$

The values of reactivity ratios can be determined by least square method. The results obtained by the symmetric equation are not spectacular, but they result in less ambiguous solution than that provided by FR method.

Kelen-Tudos Method

Kelen and Tudos underlined once again the importance of placing the Fineman-Ross equation into symmetric form. A refinement of the linearisation method was introduced by Kelen and Tudos by adding an arbitrary positive constant 'α' into the Fineman and Ross equation. This technique spreads the data more evenly over the entire composition range to produce equal weightage to all the data. The Kelen and Tudos refined form of the copolymer equation below is as follows:

$$\eta = [r_1 + r_2 / \alpha]\xi - r_2 / \alpha$$

where

$$\eta = G / (\alpha + H) \ and \ \xi = H / (\alpha + H)$$

By plotting η versus ξ, a straight line is produced that gives $-r_2/\alpha$ and r_1 as the intercepts on extrapolation to ξ=0 and ξ=1, respectively. Distribution of the experimental data symmetrically on the plot is performed by choosing the α value to be $(H_m \cdot H_M)^{1/2}$ where Hm and HM are the lowest and highest H values, respectively. The KT method has been widely used not only for free-radical copolymerisation but also for ionic copolymerisation.

Nonlinear Methods

In order to use the Mayo-Lewis equation, one must avoid the linearisation which can leads to incorrect parameter estimation. This shortcoming can lead to its "pseudo-quantitative" character.

Curve–Fitting Method

The curve-fitting method has a nonlinear form of Mayo-Lewis equation in the form of copolymer composition equation below:

$$m_1 = \frac{M_1^2(r_1 - 1) + M_1}{M_1^2(r_1 + r_2)2M_1(1 - r_2) + r_2}$$

This equation is based on the assumptions that the monomer concentrations do not change much throughout the reaction and the molecular weight of the resulting polymer is relatively high. In order to determine reactivity ratios from the experimental data, a graph must be generated for the comonomer amount incorporated into the copolymer, m_1, versus the feed comonomer amount, M_1, for the entire range of comonomer concentration. Then, a curve can be drawn through the points for selected r_1 and r_2 values and the validity of the chosen reactivity ratio values can be checked by changing the r_1 and r_2 values until the experimenter can demonstrate that the curve best fits the data points.

Braun, Brendlein, and Mott'o (BBM) attempted to make it less subjective by determining copolymer composition by r_2 at low concentration of M_1 and r_1 at high concentrations of M_1. This method has advantage of yielding reactivity ratios without the influence of the researcher and requires no estimation of the r_{ij} values in advance.

Nonlinear Tidwell Mortimer Method

Between 1964 and 1970 several fundamental papers were published which mainly emphasised nonlinear methods of determining the reactivity ratios. The nonlinear method first put forward by Behenken and then by Tidwell and Mortimer is an improved version of the curve-fitting method. Tidwell and Mortimer have pointed out the defects of the different linear methods and suggested the use of a nonlinear least squares (NLLS) procedure. Hill et al. also noted the importance of using NLLS techniques for reactivity ratios and the drawbacks of linearisation methods. A nonlinear least-squares method is the correct way to analyse copolymer composition data and to determine the reactivity ratios.

The nonlinear least-squares method is most suitable if following conditions are satisfied. (1) The terminal copolymerisation model is consistent with the experimental data; (2) errors in the dependent variable m_1 are random and statistically independent from observation to observation with constant variance; and (3) the independent variable M_1 contains no measurement error.

The Tidwell and Mortimer (TM) method employs the nonlinear least squares procedure to estimate the reactivity ratios. Briefly, the method consists of the following: given the initial estimates of r_1 and r_2 a set of computations is performed which on repetition rapidly leads to a pair of values of the reactivity ratios that yields the minimum value of the sum of the squares of the differences between the observed and computed polymer composition.

The Tidwell and Mortimer nonlinear least-squares procedure is considered to be the only statistically accurate means of determining reactivity ratios from data obtained at low conversion. The reactivity ratio values obtained by this procedure are most probable values of the system, and the TM joint confidence interval is the statistically correct one.

This method has a disadvantage. If the initial estimates are quite different from the actual values of r_1 and r_2, the value of the sum of square of deviations does not reach a minimum. In order to get rid of this, Gauss Newton nonlinear least square procedure was subjected to modifications as suggested by Box. If the initial estimates are very good then the number of iterations required for the value of the sum of square of deviations to converge is less.

The minimisation of the F criterion is used in the PROCOP programs in order to identify the optimal reactivity ratios. The PROCOP and TM method lead to the same results as both methods started with the Mayo-Lewis equation in its differential form.

Table: Comparison of the FR, KT, and TM methods for Several Copolymerisation systems.

Monomer 1	Monomer 2	r_1	r_2	FFR x 10^3	FKT x 10^3	FTM x 10^3
2,5-Dichlorostyrene	Methyl methacrylate	2.3736	0.4019	29.8	47.0	27.5

Vinylidene chloride	Styrene	2.1406	0.1249	27.1	39.8	26.9
Methyl isopropenyl ketone	Vinylidene chloride	4.0281	0.1376	22.5	160.8	14.6
α-Phenyl vinyl acetate	Butadiene	0.3300	0.2266	60.9	40.2	39.7
Styrene	Acrylonitrile	0.3471	0.0472	17.6	8.3	8.2
α-Choloro-acrolein	Styrene	0.1588	0.0245	26.5	19.7	16.0

Table compares the copolymerisation results obtained from the FR, KT and TM methods with the PROCOP program. Systems showing similar reactivity ratios were chosen, together with systems with different comonomer reactivity ratios. The table shows that in some cases the KT method better accounts for the experimental data, while in other cases the FR method yields better results. In both the situations best results are obtained through the TM method.

Kuo-Chen Method

The Kuo-Chen method is a simple nonlinear extrapolation method derived from Fineman-Ross Method. Equation is as follows:

$$\frac{M_1}{m_1} = \frac{r_1 M_1^2 + 2M_1 M_2 + r_2 M_2^2}{r_1 M_1 + M_2}$$

where M_1 and M_2 are molar concentrations of the monomers in feed and m1 and m2 are copolymer compositions. By plotting the ratio $M = M_2/m_1 m_2$ against M_1, the extrapolation to $M_1 = 0$ and $M_1 = 1.0$, respectively, gives r_1 and r_2. The challenge in using this procedure is extrapolating a nonlinear curve from a limited number of experimental points.

The KC method leads to deviation in the results, which are calculated with the TM method (PRO-COP) particularly in the case where there is significant difference in values of comonomer reactivity ratio. The reactivity ratios values estimated by the KC method are within the confidence region of copolymerisation of comonomers with similar reactivities, e.g. methyl methacrylate (M_1) and isoprene (M_2) $r_1 = 0.2320$ and $r_2 = 0.8161$.

Error in Variable Method (EVM)

A correct statistical approach to the estimation of reactivity ratios can be achieved when a method takes into account errors involved in the values of the variables (comonomer feed composition, copolymer composition, including conversion) and assesses the experimental data. The error- invariable method (EVM) is an extension of nonlinear TM method, which correctly accounts for the error in the variables. Using a principle of least squares, the sum of the weighted squares of the residuals is made a minimum. To estimate reactivity ratios using the nonlinear least square error in variable method, initial estimates of reactivity ratios are necessary and these can be obtained by KT method. In the calculation of reactivity ratios, the weights that should be assigned to an experimental point are proportional to derivatives of the response, with respect to the parameters evaluated under experimental conditions used to generate experimental points.

Optimisation Methods

The probability maximum through the minimisation of the sum of squares (SS) can be made via an optimisation method, which leading to results obtained using the TM method. If there is a small error in the values of the monomer feed composition, then it is enough to consider an error in the copolymer compositions when determining the reactivity ratios. The optimisation method developed by using PROCOP programs not only leads to results which are as good as those are obtained by TM method but also represents a more flexible method that can be used for the integral form of Mayo-Lewis equation.

As indicated earlier, to estimate the reactivity ratios with the optimisation method, initial estimates of the reactivity ratio are required. They can be determined with methods such as Fineman-Ross (FR), or Kelen-Tudos (KT) or by the intersection methods.

Methods to Estimate Reactivity Ratios at Higher Conversions

Since composition of the copolymer changes with conversion, the copolymer should be isolated at a, even at low conversions a certain composition change is recorded, and what is measured is the average molar copolymer composition, not the instantaneous composition.

Montgomery and Fry showed that the classical methods of determining monomer reactivity ratios from the differential copolymer composition equation are erroneous. The low molecular weight species formed by termination through side reaction and impurities and the handling of small quantities of copolymer magnify the errors. Purification by dissolution and precipitation entails the loss of low molecular weight species. Polymer conversions are also generally high, violating the low conversion requirement. The differential form of copolymer composition equation does not take into consideration the drift of the monomer feed, due to unequal reactivities of the monomers, with conversion. The copolymers obtained at high conversion are the true representatives of the reaction dictated by the monomer reactivity ratios, while the initial copolymers are governed to a large extent by the monomer feed.

The integrated copolymer composition equation is represented as:

$$r_2 = \frac{\log\left[\dfrac{M_2}{m_2}\right] - \dfrac{1}{P}\log\left[\dfrac{1-P(M_1/M_2)}{1-P(m_1/m_2)}\right]}{\log\left[\dfrac{M_1}{m_1}\right] + \dfrac{1}{P}\log\left[\dfrac{1-P(M_1/M_2)}{1-P(m_1/m_2)}\right]}$$

where M_1 and M_2 are the mole fractions of monomer 1 and 2 present initially and m_1 and m_2 mole fractions of the monomer remaining unreacted when the reaction is stopped. P is the integration variable, expressed as the function of the reactivity ratios:

$$P = \frac{(1-r_1)}{(1-r_2)}$$

Extended Kelen-tudos Method

Kelen and Tudos modified their low conversion equation for high conversion data by redefining η and ξ using partial molar conversion of the monomers. This is probably the most useful and widely accepted method presently. If copolymerisations are carried to higher conversions the determination of copolymerisation parameters involves labourious calculations because the integrated equation must be applied. The extended Kelen-Tudos method is used to estimate reactivity ratios with data even at high conversions. The equation may be derived by computing the partial molar monomer conversions, ζ_1 and ζ_2 and the integral z, where:

$$\varsigma_2 = w\left[\frac{\mu + x_0}{\mu + y}\right] \text{ and } \varsigma_1 = \varsigma_2 y / x_0$$

$$z = \frac{\log(1-\varsigma_1)}{\log(1-\varsigma_2)}$$

where x_0 is M_1/M_2, the initial mole ratio in feed; y is m_1/m_2, the final mole ratio in the copolymer; w is weight fraction conversion; and μ = Mol. Wt. of M_2 / Mol. Wt. of M_1. The variable H and G and the plotting parameters η and ξ can be computed as:

$$H = y/z^2 \quad and \quad G = (y-1)/z$$

$$\eta = G/(\alpha + H) \quad and \quad \xi = H/(\alpha + H)$$

The final equation for extended Kelen Tudos method is represented as:

$$\eta = [r_1 + r_2/\alpha] \ \xi - r_2/\alpha$$

where, $\alpha = (H_{min} \times H_{max})^{1/2}$. By plotting η versus ξ, a straight line is produced that gives $-r_2/\alpha$ and r_1 as the intercepts on extrapolation to $\xi=0$ and $\xi=1$, respectively. The extended Kelen-Tudos method can be applicable[67] to systems with conversions not exceeding 40%. The comparison of extended KT method with other integral methods has led to acceptable results.

The reactivity ratios can be estimated by fitting the data of the overall conversion and the cumulative copolymer composition. Kuo and Chen have put forward a very simple integrated form of the copolymer composition equation:

$$\frac{M_1 - M_1^R}{M_2 - M_2^R} = \frac{M_1(r_1 M_1^R + M_2^R)}{M_2(M_1^R + r_2 M_2^R)}$$

where, M_i^R refers to the residual molar concentration of the comonomer.

Kuo-Chen used the molar fractions for the instantaneous comonomer composition (M_i) and the cumulative copolymer composition equation, which is similar to the equation proposed by the Fineman and Ross.

Mao-huglin Method

Mao and Huglin in 1993 proposed an iterative linear method based on KT method which can be applicable to the copolymerisation system with the conversions greater than 40%. By considering the corresponding equations, computer simulation can be used for a set of data points at low and high conversions. A small number of copolymerisation systems can be placed in simplified conditions to study high conversions. This method gives a solution for simultaneous estimation of both reactivity ratios for feed composition range large enough to generate results applicable over entire range of comonomer compositions.

In this method corresponding instantaneous monomer feed composition f^c may be calculated back from the calculated value, F^c of the copolymer composition.

$$f^c = \frac{(F^c - 1) + \sqrt{(1 + F^c)^2 + 4r_1 r_2 F^c}}{2r_1}$$

F^c can be calculated from the assumed reactivity ratios by the integrated copolymerisation equation. The convergence criterion of iteration is that the recalculated reactivity ratios become equal to the assumed ones.

In the absence of experimental data on copolymerisation at higher conversion values, all that remains is to compare actual results with those of the MH method. Other investigators have previously employed many of the aforementioned linearisation methods, but the approach which is taken here, to estimate more reliable and statistically sound monomer reactivity ratios, is different.

In the present work reactivity ratios of all monomer combinations at any conversion as well as 95% joint confidence intervals were calculated using the terminal model computer program FORTRAN 77, kindly offered by Dr. R. Mao.

Intersection Method

Reactivity ratios at significantly higher conversion level were initially estimated by Mayo and Lewis. This was done through an intersection method and the integral equation $\frac{M_1}{m_1} = \frac{r_1 M_1^2 + 2M_1 M_2 + r_2 M_2^2}{r_1 M_1 + M_2}$. Monomer reactivity ratios can be estimated by determining the straight lines in an r_1-r_2 diagram for each experiment.

Binary polymerisations have become increasingly important technologically and a good knowledge of polymerisation parameters, including reactivity ratios, is very essential. Our research also took into account potential enhancements in reactivity ratio estimation for binary systems. Another related issue in multicomponent polymerisations is the existence of an azeotropic point. The feed composition of such a point would result in copolymer products with a homogeneous composition. Predicting the existence and also calculating the composition of the azeotropic point can reduce the effort of running costly experiments, in that computational results can be used to narrow the experimental search space.

Improvements suggested for the intersection method in its differential form can also be applied to its integral version. Using the F criterion allows for the selection of appoint and even the generation of new points via different forms of averaging. The application of this new approach to improve the intersection method has led to giving a point which is located very close to the optimum point.

Optimisation Method

The optimisation methods are also taken into consideration of the conversion values of system. For the ith experimental point, the instantaneous monomer feed composition M_i, the instantaneous copolymer composition mi, and overall copolymer composition gc_i can be calculated.

The aim is to find out copolymer composition mi cal starting from a monomer feed composition M_i, after integration up to the experimental conversion resulting in a cumulative copolymer composition. This copolymer composition is then compared with the experimentally determined value.

The simple algorithm is used to minimise the difference $(m_1 {}^{exp}-m_1 {}^{cal})^{76}$ for the composition of unreacted monomer. Generally an integral method (PROCOP) has been used in the optimisation method.

Error-in-Variables Method

The computational procedures have been shown that the use of an integrated form of the terminal model equation also takes into account errors in both copolymer and comonomer feed compositions. An iterative method in the computer program takes into account the effect of conversion on the polymer composition. The error-invariables method (EVM) correctly accounts for the error in the variables. The computer program accounts for the extent of conversion and the copolymer composition equation is expressed in the form consistent with the basic assumption of the least-squares procedure. The EVM takes into the account the errors observed in both independent (monomer feed) and dependent variables (the copolymer compositions).

5

Polymer Crystallization

The process by which a chemical is converted from a liquid solution into a solid crystalline state is known as crystallization. Polymer crystallization refers to a process which is associated with the partial alignment of the molecular chains in polymers. The topics elaborated in this chapter will help in gaining a better perspective about various aspects of polymer crystallization as well as its kinetics.

Crystal

Crystal is any solid material in which the component atoms are arranged in a definite pattern and whose surface regularity reflects its internal symmetry.

Classification

The definition of a solid appears obvious; a solid is generally thought of as being hard and firm. Upon inspection, however, the definition becomes less straightforward. A cube of butter, for example, is hard after being stored in a refrigerator and is clearly a solid. After remaining on the kitchen counter for a day, the same cube becomes quite soft, and it is unclear if the butter should still be considered a solid. Many crystals behave like butter in that they are hard at low temperatures but soft at higher temperatures. They are called solids at all temperatures below their melting point. A possible definition of a solid is an object that retains its shape if left undisturbed. The pertinent issue is how long the object keeps its shape. A highly viscous fluid retains its shape for an hour but not a year. A solid must keep its shape longer than that.

Basic Units of Solids

The basic units of solids are either atoms or atoms that have combined into molecules. The electrons of an atom move in orbits that form a shell structure around the nucleus. The shells are filled in a systematic order, with each shell accommodating only a small number of electrons. Different atoms have different numbers of electrons, which are distributed in a characteristic electronic structure of filled and partially filled shells. The arrangement of an atom's electrons determines its chemical properties. The properties of solids are usually predictable from the properties of their constituent atoms and molecules, and the different shell structures of atoms are therefore responsible for the diversity of solids.

All occupied shells of the argon (Ar) atom, for example, are filled, resulting in a spherical atomic shape. In solid argon the atoms are arranged according to the closest packing of these spheres. The iron (Fe) atom, in contrast, has one electron shell that is only partially filled, giving the atom a net magnetic moment. Thus, crystalline iron is a magnet. The covalent bond between two carbon (C) atoms is the strongest bond found in nature. This strong bond is responsible for making diamond the hardest solid.

Long- and Short-Range Order

A solid is crystalline if it has long-range order. Once the positions of an atom and its neighbours are known at one point, the place of each atom is known precisely throughout the crystal. Most liquids lack long-range order, although many have short-range order. Short range is defined as the first- or second-nearest neighbours of an atom. In many liquids the first-neighbour atoms are arranged in the same structure as in the corresponding solid phase. At distances that are many atoms away, however, the positions of the atoms become uncorrelated. These fluids, such as water, have short-range order but lack long-range order. Certain liquids may have short-range order in one direction and long-range order in another direction; these special substances are called liquid crystals. Solid crystals have both short-range order and long-range order.

Solids that have short-range order but lack long-range order are called amorphous. Almost any material can be made amorphous by rapid solidification from the melt (molten state). This condition is unstable, and the solid will crystallize in time. If the timescale for crystallization is years, then the amorphous state appears stable. Glasses are an example of amorphous solids. In crystalline silicon (Si) each atom is tetrahedrally bonded to four neighbours. In amorphous silicon (a-Si) the same short-range order exists, but the bond directions become changed at distances farther away from any atom. Amorphous silicon is a type of glass. Quasi crystals are another type of solid that lack long-range order.

Most solid materials found in nature exist in polycrystalline form rather than as a single crystal. They are actually composed of millions of grains (small crystals) packed together to fill all space. Each individual grain has a different orientation than its neighbours. Although long-range order exists within one grain, at the boundary between grains, the ordering changes direction. A typical piece of iron or copper (Cu) is polycrystalline. Single crystals of metals are soft and malleable, while polycrystalline metals are harder and stronger and are more useful industrially. Most polycrystalline materials can be made into large single crystals after extended heat treatment. In the past blacksmiths would heat a piece of metal to make it malleable: heat makes a few grains grow large by incorporating smaller ones. The smiths would bend the softened metal into shape and then pound it awhile; the pounding would make it polycrystalline again, increasing its strength.

Categories of Crystals

Crystals are classified in general categories, such as insulators, metals, semiconductors, and molecular solids. A single crystal of an insulator is usually transparent and resembles a piece of glass. Metals are shiny unless they have rusted. Semiconductors are sometimes shiny and sometimes transparent but are never rusty. Many crystals can be classified as a single type of solid, while others have intermediate behaviour. Cadmium sulfide (CdS) can be prepared in pure form and is an excellent insulator; when impurities are added to cadmium sulfide, it becomes an interesting semiconductor. Bismuth (Bi) appears to be a metal, but the number of electrons available for electrical conduction is similar to that of semiconductors. In fact, bismuth is called a semimetal.

Molecular solids are usually crystals formed from molecules or polymers. They can be insulating, semiconducting, or metallic, depending on the type of molecules in the crystal. New molecules are continuously being synthesized, and many are made into crystals. The number of different crystals is enormous.

Crystallization

Crystallization is the solidification of atoms or molecules into a highly structured form called a crystal. Usually, this refers to the slow precipitation of crystals from a solution of a substance. However, crystals can form from a pure melt or directly from deposition from the gas phase. Crystallization can also refer to the solid-liquid separation and purification technique in which mass transfer occurs from the liquid solution to a pure solid crystalline phase.

Detached retina imaging

Although crystallization may occur during precipitation, the two terms are not interchangeable. Precipitation simply refers to the formation of an insoluble (solid) from a chemical reaction. A precipitate may be amorphous or crystalline.

Process of Crystallization

Two events must occur for crystallization to occur. First, atoms or molecules cluster together on the microscopic scale in a process called *nucleation*. If the clusters become stable and sufficiently large, *crystal growth* may occur. Atoms and compounds can generally form more than one crystal structure (polymorphism). The arrangement of particles is determined during the nucleation stage of crystallization. This may be influenced by multiple factors, including temperature, the concentration of the particles, pressure, and the purity of the material.

In a solution in the crystal growth phase, an equilibrium is established in which solute particles dissolve back into the solution and precipitate as a solid. If the solution is supersaturated, this drives crystallization because the solvent cannot support continued dissolving. Sometimes having a supersaturated solution is insufficient to induce crystallization. It may be necessary to provide a seed crystal or a rough surface to start nucleation and growth.

Examples of Crystallization

A material may crystallize either naturally or artificially and either quickly or over geological timescales. Examples of natural crystallization include:

- Snowflake formation.
- Crystallization of honey in a jar.
- Stalactite and stalagmite formation.
- Gemstone crystal deposition.

Examples of artificial crystallization include:

- Growing sugar crystals in a jar.
- Production of synthetic gemstones.

Crystallization Methods

There are many methods used to crystallize a substance. To a large degree, these depend on whether the starting material is an ionic compound (e.g., salt), covalent compound (e.g., sugar or menthol), or a metal (e.g., silver or steel). Ways of growing crystals include:

- Cooling a solution or melt.
- Evaporating solvent.
- Adding a second solvent to reduce the solubility of the solute.
- Sublimation.
- Solvent layering.
- Adding another cation or anion.

The most common process is to dissolve the solute in a solvent in which it is at least partially soluble. Often the temperature of the solution is increased to increase solubility so the maximum amount of solute goes into solution. Next, the warm or hot mixture is filtered to remove undissolved material or impurities. The remaining solution (the filtrate) is allowed to slowly cool to induce crystallization. The crystals may be removed from the solution and allowed to dry or else washed using a solvent in which they are insoluble. If the process is repeated to increase the purity of the sample, it is called recrystallization.

The rate of cooling of the solution and the amount of evaporation of the solvent can greatly impact the size and shape of the resulting crystals. Generally, slower is better: slowly cool the solution and minimize evaporation.

Fundamentals of Crystallization

How does the Crystallization Process Occur?

The crystallization process consists of two major events:

- Nucleation: Molecules gather together in clusters in a defined manner. Clusters need to be stable under current experimental conditions to reach the "critical cluster size" or they will redissolve. It is this point in the crystallization process that defines the crystal structure.

- Crystal Growth: Nuclei that have successfully achieved the "critical cluster size" begin to increase in size. Crystal growth is a dynamic process, with atoms precipitating from solution and becoming redissolved. Supersaturation and supercooling are two of the most common driving forces behind crystal formation.

Development of crystallization processes represents a complex and challenging issue, requiring simultaneous control of various product properties, including purity, crystal size and shape, and molecular level solid structure. The control of the nucleation phase is difficult but is the key to process control; crystallization chemists usually aim to achieve goals of high purity and high yield by solely using controlled cooling crystallization techniques.

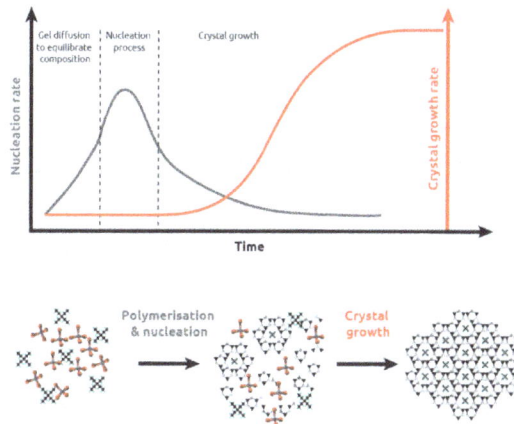

The crystallization process - crystal growth rate vs. nucleation rate.

Many compounds can exist in multiple crystal structures – a phenomenon known as "polymorphism" – and can have different physical properties (melting point, shape, dissolution rate, etc). Depending on the conditions used, either nucleation or crystal growth may be predominant over the other, leading to crystals with different shapes and sizes. Therefore, controlling polymorphism is of significant interest in chemical manufacture.

A common example of the importance of crystal size can be found with ice-cream. Small ice crystals, formed through rapid cooling, improve the texture and taste of the ice-cream compared with larger ice crystals.

Solubility Curves, Supersaturation and the Metastable Zone Width

Traditionally, crystal formation has been achieved by reducing the solubility of the solute in a saturated solution in a variety of ways.

Solubility curves are a common tool used by scientists to understand/demonstrate the relationship between solubility, temperature, and solvent type. By plotting this information, scientists can find the optimum solvent/anti-solvent, temperature, and theoretical yield for their crystallization process.

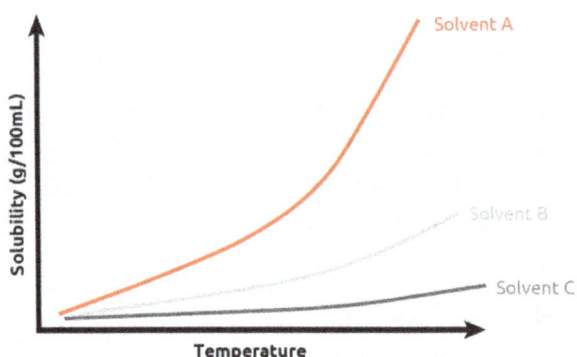

A solubility and temperature graph for various solvents

In figure shows that the given material is highly soluble in Solvent A, meaning more material can be crystallized from a given volume of solvent. Conversely, the given material has a low solubility in Solvent C across all temperatures, potentially making it a good anti-solvent for this material.

Calculating Theoretical Crystallization Yield

The theoretical yield of crystallization can be calculated at various temperatures: If a saturated solution containing 45 g of product per 100 g solvent A is cooled from 50 °C to 20 °C, then 15 g of product per 100 g of solvent will remain in solution. Therefore, 30 g of product should crystallize, allowing scientists to measure the yield/efficiency of their crystallization.

In reality, when a saturated solution is cooled, there is more solute in the solution than predicted by the solubility curve, and this is referred to as "supersaturation".

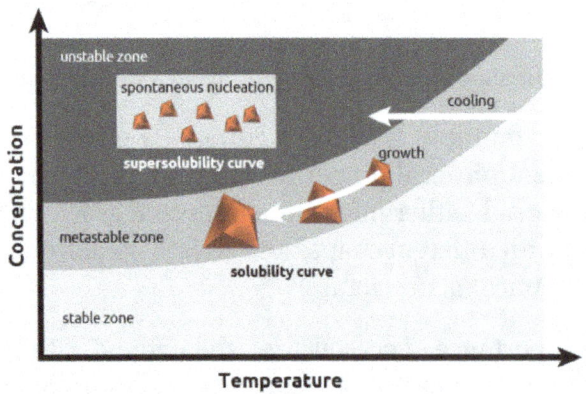

Crystallization solubility curve.

As cooling continues, at a certain temperature, crystal nucleation will begin. This temperature is called the "metastable limit", and the difference between this temperature and the solubility curve is known as the metastable zone width (MSZW).

By carefully controlling the level of supersaturation of a solution, scientists can control the crystallization process.

As can be seen from the above schematic, at low levels of supersaturation, crystals grow more quickly than they nucleate resulting in large crystal size distribution. At high supersaturation levels, nucleation dominates crystal growth, providing smaller crystals. This makes understanding and controlling supersaturation vitally important when creating crystals of a desired size and distribution.

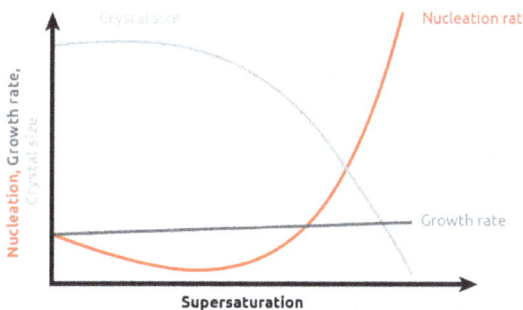

Nucleation, growth rate, and crystal size.

Role of Crystallization Play in Industrial Research, Development and Manufacturing

Crystallization is one of the most widely used technologies in chemical industry, and process robustness governs process productivity and economics. In particular, the pharmaceutical and food sectors are utilizing crystallization for optimized separation, purification, and solid form selection. For example, crystallization is the most common method of formation of pharmaceutical solids for Active Pharmaceutical Ingredient (API) development. The optimization of the particulate properties such as particle size and shape distributions is paramount as the physical form dictates drug product quality and effectiveness.

Many pharmaceutical drugs have poor physiochemical profiles, such as poor solubility in biological fluids. Significant research and development efforts have been made towards developing a solid form landscape that covers all possible solid structures, including polymorphs, solvates, co-crystals, salts, and the amorphous phase to improve Active Pharmaceutical Ingredient (API) development.

Methods of Crystallization

Crystallization is the oldest "unit operation" in a chemical engineering sense. For example, Sodium Chloride has been manufactured this way since the dawn of civilization.

Various traditional methods for crystallization exist, with each technique having unique benefits and drawbacks. The method chosen must be selected based on the properties of the material being crystallized.

- Solvent Evaporation: Easy to set-up, requires air stable samples, requires a minimum solvent volume to work effectively. Large amount of material required.

- Slow Cooling: Requires solvents with boiling points less than 100 °C and moderate solute solubility. Large amount of material required.

- Solvent/Vapour Diffusion: Works well with small amounts of material, however finding two suitable solvents can be challenging. Can "oil out".

- Sublimation: Not the method of choice for diffraction quality crystals. Typically performed at high temperatures, causing crystals to grow too quickly.

Sonocrystallization

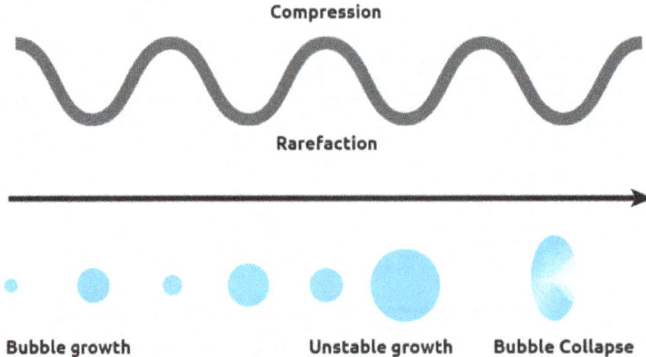

The sonocrystallization process - compression and rarefaction.

Crystallization processes are often difficult to control, but sonocrystallization is a more modern method of crystallization that offers significant advantages over traditional methods. Ultrasound radiation is known to induce acoustic cavitation in liquids through the formation, growth, and collapse of bubbles. The collapse of the bubble provides energy to encourage the nucleation process at the earliest possible point in time. This results in highly repeatable and predictable crystallizations, and offers various benefits, including:

- Reduced induction time.
- A decrease in metastable zone width (MSZW).
- Increased nucleation rate.
- Increased crystal growth rate.
- Reduce agglomeration.
- Tailored crystal size distribution.

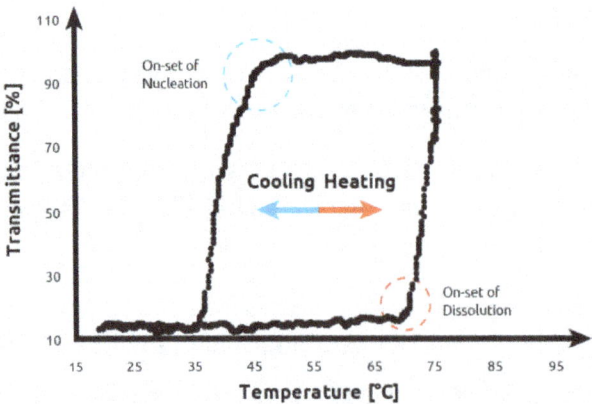

Onset of nucleation and dissolution graph in the sonocrystallization process.

Crystallization Monitoring Methods

Having decided on a method of crystallization, it is important to measure the progress and subsequent success of your crystallization process.

Measuring Crystallization using Turbidity

Turbidity probes have been used to monitor crystallizations for decades, due to their ease of use, sensitivity, and affordability. Turbidity probes work by measuring light that is scattered by suspended solids in a liquid. As the total suspended solids increases, the turbidity level (and the cloudiness/haziness) increases, making it a useful tool for calculating the metastable zone of a system and monitoring the formation of crystals.

To create a solubility curve, different concentrations of solute are ramped up and down in temperature, allowing nucleation/dissolution points to be calculated. As an example, turbidity information for the crystallization and dissolution of adipic acid is shown below:

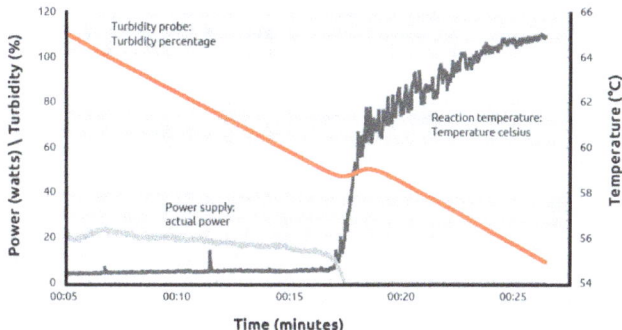

As can be seen on the graph, at a given concentration, adipic acid will spontaneously nucleate at approximately 59 °C and the solution clears at approximately 64 °C giving a metastable zone width of 5 °C.

Melting of Crystals

Melting, the fundamental process in which, at a certain temperature T_m, a crystalline substance undergoes a phase change from a solid to a liquid (melt) still holds mysteries despite its common occurrence. One reason is that in experiments it is still not feasible to observe directly the atomistic details of the process. This means that we do not know the structural arrangements of the atoms or their characteristic motions prior to and during melting-information which is needed in formulating a fundamental theory of the transition. Another reason is that most theoretical methods, including atomic-level computer simulations, do not take into account the effects of extrinsic lattice defects, such as surfaces, dislocations and grain boundaries. The role of lattice defects in the onset of the destruction of long-range order has, consequently, not been clearly established.

Our inability to see how melting occurs does not prevent us from knowing why it occurs and when the order-disorder transition should take place. According to thermodynamics, the melting point T_m is that temperature at which the solid and liquid phases can coexist in equilibrium, a condition which occurs when the Gibbs free energies of the two phases are equal. It is implied that at temperatures above this coexistence temperature the crystal is unstable. However, thermodynamics says nothing about what the mechanism of melting is or how long the process will take. These questions are related to the kinetics of the phenomenon. For a more complete understanding of melting one therefore needs to be concerned with both thermodynamics and kinetics.

Theories of Melting

In a general discussion of melting, one should distinguish from the outset between intrinsic and extrinsic lattice defects. Intrinsic defects, such as lattice vacancies, are produced thermally. By contrast, due to the increase in free energy associated with their creation, extrinsic defects are usually thermodynamically metastable. For example, grain boundaries can be eliminated from polycrystals by high-temperature annealing, thus allowing the material to achieve a state of lower free energy during the process of recrystallization. A number of theories of melting have been proposed during the past seventy years or so, all of which consider only the effects of intrinsic defects. These theories generally assume the dominant mechanism to be one of three types:

a) According to Lindemann, melting is caused by the onset of an instability when the displacements during thermal vibration of the atoms exceed a certain threshold value ("Lindemann criterion").

b) According to Born, melting arises from the onset of a mechanical instability, manifesting itself in an imaginary phonon frequency at the onset of which the crystal lattice collapses ("Born instability").

c) In the theories of Cahn and others, the spontaneous production of intrinsic lattice defects, such as vacancies and intrinsic arrays of dislocations near the melting point, is thought to be responsible for the breakdown of long-range crystalline order.

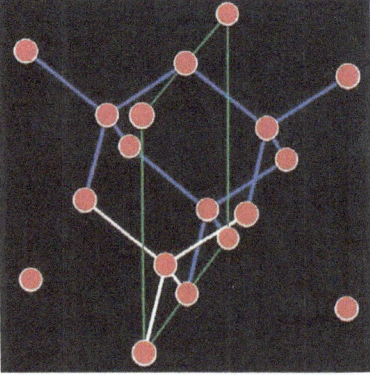

The above figure shows Crystallographic unit cell of the diamond lattice, with silicon atoms denoted by red balls and nearestneighbor bonds signifiedin blue. The bonds of one of the four-fold coordinated silicon atoms are highlighted in white. The green rectangle illustrates a (Ito) plane. The edge of this cubic unit cell defines the lattice parameter, a = 5.431Å, of Si. (1 Angstrom = to10⁻¹⁰m).

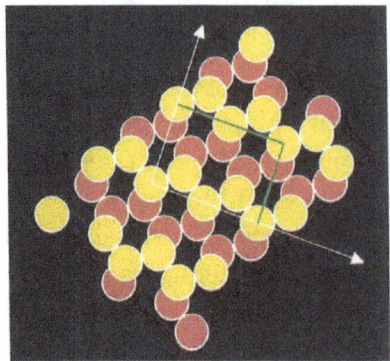

The above figure shows View of the computational cell from above showing the top two (110) planes each containing 22 Si atoms as yellow and red balls, respectively. The dotted gold line encloses the computational cell, which is rotated by 25.24° relative to the < 110> and < 100> directions in the crystal; the latter are indicated by white arrows. The smaller area enclosed by the green and white lines corresponds to the fraction contained in the single cubic unit cell.

These theories have in common that melting is considered as a process occurring homogeneously throughout the crystal, with the effects of surfaces both internal and external, being entirely neglected.

Experimentally, however, it is well known that melting generally proceeds from surfaces into the interior crystalline regions-a process requiring a finite amount of time. For example, about 30 years ago Turnbull and coworkers demonstrated that melting of silica, SiO_2 and phosphorous pentoxide, P_2O_5, are not homogeneous processes but that, instead, the liquid phase nucleates at free surfaces and grain boundaries, from which it propagates into the crystal. Also, recent laboratory experiments by Daeges, Gleiter and Perepezko have demonstrated that small single crystals of silver can be superheated above the melting temperature when coated with gold (which has an almost identical lattice parameter but a higher melting point). Such a coating replaces the silver free surface with a silver-gold interface. Because the lattice parameters of the materials are so similar. However, the effects of such an interface should be small. It was found that the coated pellets could be heated well past the melting point of silver whereas normally, superheating above T_m is known to be difficult to achieve in metals. These observations not only raise the question of whether melting can, indeed, be regarded as a homogeneous process, but also point to the need for a direct simulational study of melting where surface effects can be isolated, and analyzed in terms of thermodynamics and kinetics.

Molecular-Dynamics Simulation of Silicon at High Temperature

Atomistic modeling, in the form of moleculardynamics (MD) and Monte Carlo simulations, is a method of studying the cooperative and individual behavior of a system of atoms under well-prescribed conditions. Through the use of interatomic interaction potentials and border conditions, simulations can be made to represent the physical state of a material at finite temperatures and pressures reasonably well. The results of such simulations are particularly valuable for the study of the relation between the atomic structure of a system and its thermodynamic, mechanical, and transport properties. Because of these features, atomistic simulation is increasingly gaining recognition as a means of probing complex physical processes at the atomic level.

The choice of silicon for our simulation of the melting process is motivated by three factors. First, due to its covalent nature of bonding, the crystal structure of silicon is that of cubic diamond, with only four nearest neighbors. Every Si atom may thus be considered to occupy the center of a tetrahedron, with the four nearest neighbors at the comers. Upon melting, the tetrahedral coordination is destroyed in favor of an average six-fold coordination in the liquid. By monitoring the average number of nearest neighbors of every atom during the simulation, the four-coordinated "crystalline" Si atoms may readily be distinguished from the six-coordinated "liquid" Si atoms. The fact that a good empirical interatomic potential is available for silicon (derived by Stillinger and Weber) which has been shown to give a rather realistic overall description of its physical properties, provides another reason for our choice of silicon. Finally, silicon is a good choice because extensive experimental data exists on its melting and freezing behavior, including the well-known fact that, like ice, silicon contracts upon melting.

In any study of melting, knowledge of the thermodynamic melting point, Tm s is of primary importance. Although any interatomic potential function permits only an approximate description of a
given material, it is essential to precisely determine T_m ' Using MD methods to determine the temperature-dependent free energies for the crystalline and liquid phases, Broughton and Li obtained
a value of Tm = 1691 ± 20K for the StillingerWeber potential. That this value is quite close to the
known melting point of silicon (1683K) indicates that both the Stillinger-Weber potential function
and the molecular-dynamics approach are valid for the study of melting.

A typical molecular-dynamics simulation cell contains several hundred or a few thousand atoms,
at most. The basic information that one obtains from such simulations includes, most important-
ly, the positions and velocities of all the atoms in the system at any instant during the simulation.
These variables are obtained by numerically solving Newton's equations of motion for a system of
many interacting atoms. Within the limitations of any simulation (pertaining to the approximate
nature of the interatomic potential used, and the finite system size and duration of the simulation),
the atom positions are like the recorded output of a hypothetical atomic camera, operating with
a field of view spanning a distance of no more than about 100 Å. at a speed of 10^{14} to 10^{15} frames
per second. On such microscopic scales, one can locate unambiguously the position of each atom
and follow its motion as it interacts with its neighbors. In the case of silicon we have chosen a time
interval of 1.15×10^{-15}s over which Newton's equations of motion are integrated in each time step
of the simulation. With lattice-vibration frequencies of typically about $lo^{13}s^{-1}$ for Si, a simulation
interval of 1000 time steps (= 1.15 ps) covers approximately 10 vibration periods of each atom.

The purpose of our simulations is to isolate the effects of surfaces and interfaces, and to elucidate
the kinetics of the melting process. To investigate the roles played by internal and external surfaces, we have carried out a series of molecular-dynamics simulations of the melting of crystalline
silicon, with and without such extrinsic defects, at elevated temperatures. Three model geometries
will be considered; the first two focus on the role of grain boundaries and free surfaces, while in
the third we investigate the melting behavior of a defect-free perfect crystal. The simulation-cell
dimensions, including the number of simulated atoms in the cell, will be chosen to be the same in
all three situations, thus eliminating any effects which might possibly arise from different dimensions of the simulation cell as well as numbers of atoms in the cell.

We first investigate the effect of the grain boundary, followed by a simulation of the free (110) surface, in the onset of melting. The planar arrangement of atoms in the rectangular unit cell on the
(110) plane is illustrated in figures. As far as the simulation of melting is concerned, this choice
of (110) planes has the advantage that their interplanar spacing, d(110) = 0.354a, is much larger
than the vibrational amplitudes of the atoms, which are typically no more than about 10% of the
nearestneighbor distance of 0.433a even at the highest temperatures.

Order Parameter for the Melting Transition

Before discussing our simulation results, an order parameter, representing a quantitative measure
for characterizing the crystalline and melted regions, has to be defined. If melting were to begin
at the surfaces or grain boundaries, then in order to distinguish crystalline from melted regions, it
would appear natural, for the purpose of analysis, to subdivide the computational cell into a finite
number of slices parallel to the surface or grain boundary and to define some order parameter, ξ,
which characterizes quantitatively the degree of crystallinity within each slice. As is common in the

area of phase transformations, ξ should be defined such that it varies within the bounds 0<ξ<1.

The static structure factor, S (k), is essentially the Fourier transform of the distribution of bond lengths, and may be considered as an order parameter of the crystalline-to-liquid phase transition. From x-ray structure determinations, S(k) is well known to characterize the long-range order in the direction of any vector, k, which is a vector of the reciprocal space lattice. The vectors of this lattice are related to the position vectors, r_i, of the atoms in the real-space lattice by the familiar Ewald relationship:

$$(k \cdot r_i) = 2\pi.$$

Being a complex function, with a real and an imaginary part, S(k) directly does not satisfy the requirements imposed on the order parameter ξ However, its square, given by

$$[S(k)]^2 = \left| S(k) \right|^2 = [1/N \sum_i \cos(k \cdot r_i)]^2 + [1/N \sum_i \sin(k \cdot r_i)]^2,$$

varies between zero and one, depending on the values of (k-r_i) which range between zero and 2π. We note that the summation in equation $[S(k)]^2 = \left| S(k) \right|^2 = [1/N \sum_i \cos(k \cdot r_i)]^2 + [1/N \sum_i \sin(k \cdot r_i)]^2$, includes all atoms i in the crystal, with k being some arbitrary vector. One readily sees that if k is, indeed, a reciprocal-lattice vector, satisfying equation $(k \cdot r_i) = 2\pi$ then equation $[S(k)]^2 = \left| S(k) \right|^2 = [1/N \sum_i \cos(k \cdot r_i)]^2 + [1/N \sum_i \sin(k \cdot r_i)]^2$, yields $[S(k)]^2 \equiv 1$. Since we are interested in the planarlong-range order parallel to the (110) planes (rather than the overall static structure in equation $[S(k)]^2 = \left| S(k) \right|^2 = [1/N \sum_i \cos(k \cdot r_i)]^2 + [1/N \sum_i \sin(k \cdot r_i)]^2$, we define a planar static structure factor.

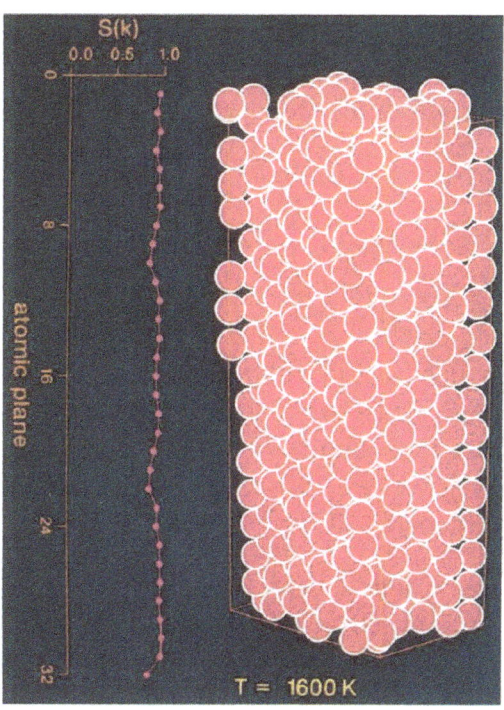

The above figure shows the Ideal-crystal computational cell of 704 atoms, stacked in 32 (110) planes, each containing 22 Si atoms. The limits of the cell are indicated by the gold box, beyond which 3-d periodic border conditions are applied to simulate an infinitely large perfect crystal. After gradually heating to 1600K (i.e., 91K below T_m for the Stillinger Weber potential), all atoms have the ideal-crystal coordination of four. Here and in subsequent figures, the color of atoms indicates their nearest-neighbor coordination K, with red, blue and green denoting, respectively, K = 4, K < 3 and K>5.

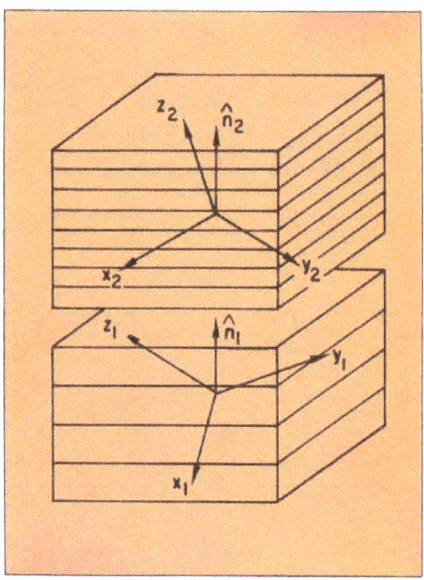

Definition of the five degrees of freedom of a general grain boundary contained in a bicrystal. Far from the boundary plane in the direction of the interface-plane normal, the inhomogeneous region containing the interface is on both sides embedded between perfect-crystal material.

The plane-by-plane profile of the order parameter, S(k), at 1600K is shown on the left for a reciprocal lattice vector, k, in the < 110> direction within each of the (110) planes. The values of S(k) near unity illustrate the large degree of crystalline long-range ordering within the (110) planes, in spite of the substantial thermal vibrations of the atoms at this very high temperature.

S_p (k), the square of which represents the desired order parameter, according to:

$$\xi \equiv [S_p(k)]^2 = [1/N\sum_{i(p)} \cos(k \cdot r_i)]^2$$
$$+ [1/N\sum_{i(p)} \sin(k \cdot r_i)]^2,$$

in which only atoms i = i(p) in a given lattice plane, p, are considered. For a perfect crystal lattice at zero temperature, t is identically equal to one for any wave vector, k, which is a reciprocal lattice vector in that plane. By contrast, in the liquid state without any long-range order in the planes, t fluctuates near zero. As a reminder of its association with the static structure factor, in the following the order parameter will be denoted simply by S (k) ($\equiv\xi$).

When the crystal is heated from absolute zero, the value of S (k) is reduced slightly from unity because of the lattice vibrations. This effect is illustrated in figure for a crystal containing a total of 704 Si atoms (arranged in 32 (110) planes, each of which contains 22 atoms in the rectangular planar unit cell; the reason for the choice of this particular unit cell will become more apparent below). In this simulation, as is common in MD simulations, three dimensional (3-d) periodic border conditions were applied to the simulation cell in order to approximate an infinitely large crystal. The crystal shown in figure above was slowly heated from zero temperature to l600K, which is 91K below the thermodynamic melting point for the Stillinger Weber potential. At this temperature all atoms are perfectly coordinated, with four nearest neighbors. The plane-by-plane order parameter, shown in the left of the figure, demonstrates a typical reduction by about 10 percent at this elevated temperature due to the lattice vibrations. The reduction in S (k) from unity is governed by the so-called Debye-Waller factor which represents a measure for the average vibrational amplitude of the atoms at a given temperature.

Role of Grain Boundaries

Grain boundaries are of considerable interest in the context of melting because most real materials are polycrystals. A grain boundary is generally characterized by five geometrical degrees of freedom (DOF) defined in above figure which illustrates the formation of a bicrystal, with the grain boundary (GB) in its center, by joining two single crystals along two well-defined crystallographic planes. We recall that any crystallographic direction, $\hat{n}$, given for example .in terms of Miller indices, <h,k,l>, represents two DOF, according to:

$$\hat{n} = (h^2 + k^2 + 1^2)^{-1/2} \begin{pmatrix} h \\ k \\ l \end{pmatrix}.$$

As illustrated in the figure above the characterization of the 5 DOF of the grain boundary starts with defining the GB-plane normal, $\hat{n}$, in each of the two principal coordinate systems, (x_1, Y_1, Z_1) and (x_2, Y_2, Z_2), associated with the two halves of the bicrystal. For example, the GB-plane normal ill in the first of the two semicrystals might be a < 110> direction while in the other crystal the normal, $\hat{n}_2$, might be a < 111> direction.

With the GB-plane normal thus fixed with respect to the two crystals (and thus having defined four ofthe 5 DOF), the only remaining DOF is the one associated with a so called twist rotation about $\hat{n}$, characterized by the twist angle θ. The 5 DOF of a general grain boundary may then be summarized as follows:

$$\{DOF\} = \{\hat{n}_1, \hat{n}_2, \theta\}.$$

In a symmetrical grain boundary, ill and il2 represent the same set of crystallographically equivalent lattice planes (i.e., $\hat{n}_1 = \hat{n}_2$), leaving only three DOF, namely $\hat{n}$ and θ.

The GB we have studied is symmetrical, with the GB plane being a (110) plane. To introduce the GB into the bicrystal, the one half is rotated by $\theta = 50.48°$ with respect to the other about the < 110> plane normal. In terms of Equation $\{DOF\} = \{\hat{n}_1, \hat{n}_2, \theta\}$, this interface is therefore characterized by $\{DOF\} = \{< 110>, < 110>, 50.48°\}$. In the jargon of the grain-boundary community, this boundary is referred to as the (110) 50.48° $(\sum 11)$ twist boundary. The value of indicates that the area of its planar unit cell is = 11 times larger than that of a perfect-crystal (110) plane in Si. The unit-cell dimensions of this GB, illustrated in figure explain our earlier choice of simulation cells for the perfect-crystal simulation: this choice renders the simulation-cell dimensions identical for the perfect crystal, the thin slab and the bicrystal, the only difference arising from the different border conditions imposed on this cell.

Similar to the above figure each of the 32 (110) planes in the bicrystal contains 22 atoms. With an interplanar spacing of (110) planes in Si of d(110) = 0.354a, the length of the computational cell is 11.33a. The planar unit-cell area of 2.345a X 3.3l7a=7.778a2 is 11 times that of the corresponding primitive planar unit cell. (Notice that the planar dimensions in both the primitive and the simulation cell are related by $\sqrt{2}$.) The computational cell thus contains 704 Si atoms in a volume of 88.l29a³.

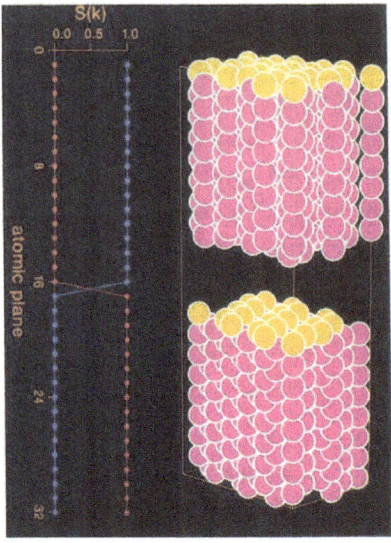

The above figure shows - (110)() = 50.48° bicrystal of Si at zero temperature. The bottom half of the bicrystal has the same orientation as the computational cell used in the perfect-crystal and thin-slab simulations. The upper half of the bicrystal is rotated by θ = 50.48° about the < 110> common grain-boundary plane normal. The atoms in the top plane of each half of the bicrystal are colored in yellow to highlight their relative rotation. For illustrative purposes, in this and in the following figure, the top and bottom halves of the bicrystal are separated, so as to make the evolution of the system more clear. The two wave vectors, k_1 and k_2, in the order parameters $S(k_1)$ and $S(k_2)$ (indicated, respectively, in red and blue in the left of the figure) represent reciprocal planar lattice vectors in the related half of the bicrystal. At zero temperature, in semicrystal 1 we then have, $S(k_1)$ = 1 and $S(k_2)$ = 0 whereas in semicrystal 2, $S(k_2)$ = 1 and $S(k_1)$ = 0.

Far from the interface, which is characterized by a 2- d periodic arrangement of atoms parallel to the GB plane, any grain boundary is surrounded by perfect-crystal material. We apply 2-d periodic border conditions in the interface plane to simulate the infinite size of the bicrystal parallel to the GB plane. However, in the direction of the GBplane normal neither the free borders of the thin film nor the periodic borders of the perfect crystal are appropriate. By surrounding the simulation cell on both sides of the grain boundary with rigid perfect-crystal blocks which are allowed to slide parallel and perpendicular to the interface plane, thus enabling both GB migration and a volume expansion at the boundary, we have recently developed a border condition which permits a realistic simulation of "bulk" interfaces, i.e., interfaces embedded between perfect material.

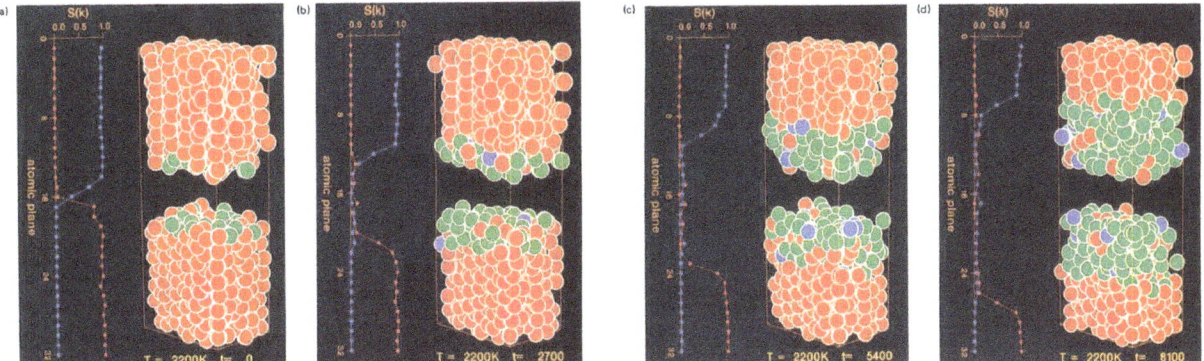

The above figure shows Heterogeneous "thermodynamic melting" of a silicon bicrystal containing in its center the (110) ($\theta = 50.48°$ ($\sum 11$) grain boundary. After having been heated over a period of 600 time steps from 1600K (T T_m), the time was reset to t = 0.1000 time steps corresponds to 1.15 psec of real time. The color of the atoms indicates their nearest neighbor coordination K, with red, blue and green denoting, respectively, K = 4, K<3 and K>5.

a) At t = 0 (i.e., immediately after the simulation temperature of 2200K was reached), there are a number of atoms at the grain boundary that already have coordination greater than four. The order parameters show, however, the sharp definition of the GB region containing only about 4 (110) planes.

b) After 2700 time steps, a number of planes on either side of the GB plane have melted. The near-zero values of the structure factor show that long-range order has now broken down in approximately seven (110) planes closest to the GB.

c) After 5400 time steps more planes have melted.

d) After 8100time steps over half of the system has melted; long-range order has been lost in the 20 central plane of the system.

In contrast to the perfect-crystal, two order parameters are now required to investigate the break-down of planar crystalline order upon melting one associated with each half of the bicrystal. By choosing two wave vectors, k_1, and k_2, which are reciprocal lattice vectors in the (110) planes associated with the two semicrystals, we define the order parameters $S(k_1)$ and $S(k_2)$ in analogy to Equation $\xi \equiv [S_p(k)]^2 = [1/N\sum_{i(p)} \cos(k\cdot r_i)]^2 + [1/N\sum_{i(p)} \sin(k\cdot r_i)]^2$. ($k_1$ and k_2 are thus simply related by the relative rotation of the two halves of the bicrystal about the < 110> GB-plane normal). By monitoring these two order parameters, every lattice plane may then be characterized as either being totally disordered (for $S(k_1) \approx S(k_2) \approx 0$), as belonging to semicrystal 1 (for $S(k_1) \approx 1$, $S(k_2) \approx 0$), or as belonging to semicrystal 2 (for $S(k_2) \approx 1$, $S(k_1) \approx 0$). A slice-by-slice orderparameter profile then shows a sudden drop at the GB in $S(k_1)$ from unity to zero (red symbols in the left of figure) , while $S(k_2)$ increases simultaneously from zero to one (blue symbols).

To investigate the high-temperature behavior of the bicrystal, the system was first equilibrated at 1600K. The temperature was then stepped up rather rapidly in intervals of 100K, allowing 100 time steps for approximate equilibration, until the desired final simulation temperature, ranging between 1800K and 2200K, was reached; this instant will be labeled t = 0. As already mentioned, the average atomic coordination increases from four to approximately six upon melting.

To illustrate the different local environments of the atoms in the bulk, a surface or in the liquid, throughout our discussion red atoms will indicate perfect-crystal coordination (K = 4), with green atoms indicating K > 4 while for blue atoms K<4.

The response of the bicrystal when heated to a temperature above T_m is shown in the four snapshots in figure below which is characteristic of all of our grain-boundary simulations. It appears that melting begins at the grain boundary, from which the melted layer then spreads into the bulk regions. The snapshot at t = 0 (i.e., immediately after reaching the desired simulation temperature) illustrates two important features. First, as indicated by the absence of blue atoms, all Si atoms near the GB are at least four-fold coordinated. The presence of a significant number of green atoms suggests that numerous atoms near the GB have already reached liquid-like coordination during the short heating period above T_m. Second, as demonstrated by the crossing of $S(k_1)$ and $S(k_2)$ in the left of figure and by the narrow region of only about four (110) planes in which the order parameters deviate significantly from unity, the GB is sharply defined, as sharply as it was prior to raising the temperature from 1600K to a temperature above T_m.

The order parameters in the left of figure above (b)-(d) clearly show the propagation of two interfaces which separate areas of well-ordered (110) planes from areas with no planar long-range order at all (and, consequently, with S(k) fluctuating near zero). A detailed analysis of the atom positions in the direction of the GB-plane normal shows the breakdown of a planar atom arrangement in the regions with vanishing order parameter.

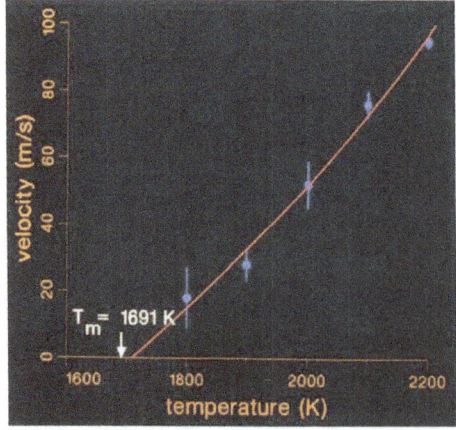

Propagation velocities of the solid-liquid interfaces as a function of temperature. The red curve, representing a quadratic fit to the data points, extrapolates to zero velocity at T = 1710 ± 30K, as compared with the thermodynamic melting temperature of the Stillinger Weber potential of Tm = 1691 ± 20K.

That the disordered region is, indeed, liquid (and not merely disordered like in an amorphous solid) was verified in two ways. First, the related volume contraction upon melting agrees well with that determined from an independent simulation of liquid silicon. Second, the plane-by-plane mean-square displacements, $< r^2 >$, of the atoms are found to increase linearly with simulation time in the disordered regions, but to be practically constant in the crystalline regions away from the GB. In MD simulations, such a linear increase in, associated with the random walk of the atoms during self-diffusion, is considered a fingerprint of a liquid. Moreover, the value of the self-diffusion constant extracted from the slope of $< r^2 >$ versus t is in good agreement with that determined from separate simulations of the liquid at the same temperature. We therefore conclude that the disordered regions are, indeed, liquid.

The above simulations reveal that melting begins at the grain boundary from which the liquid phase penetrates into the crystalline region via the propagation of two solid liquid interfaces. To actually follow the movement of these interfaces requires a certain amount of data analysis because the atoms move continuously, and the position of the interface is not always easy to determine. Nevertheless, from the results of our simulations at several temperatures above Tm s propagation velocities for the spreading of the solid-liquid interfaces into the crystalline regions, v, can be extracted. In figure these velocities are plotted as function of temperature. Extrapolation of these temperature-dependent velocities to zero velocity should yield an estimate of the coexistence temperature, T_m, at which the crystal and liquid are in thermodynamic equilibrium, and at which, therefore, the solid-liquid interface will not propagate.

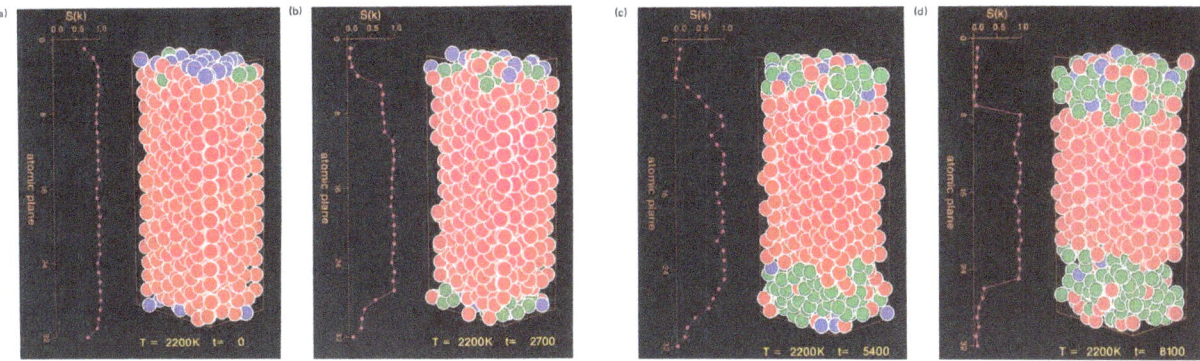

The above figure shows the Evolution with time of a thin slab with (110) faces after having been heated over a period of 600 time steps from 1600K ($T < T_m$), at which instant the time was reset to t = 0.1000 time steps correspond to 1.15 p sec of real time. The color of the atoms indicates their nearest-neighbor coordination K, with red, blue and green denoting, respectively, K = 4, K 5.

a) At t = 0 most atoms on the surfaces are three-fold coordinated (blue balls), due to the removal of one nearest neighbor on creation of the free surface. A few surface atoms have already become greater than four-fold coordinated during the 500 time steps during which the system was at $T > T_m$.

b) After 2700 time steps more atoms are liquid-like coordinated (green balls), although the liquid region has hardly spread into the bulk. Note however, that the structure factor shows the breakdown of in-plane long range order for a number of planes away from the surfaces.

c) After 5400 time steps there are significant regions of melted material near the two free surfaces. As indicated in the order parameter, a thermal fluctuation has apparently caused a small amount of recovery in the long range order of the system near the lower free surface.

(d) After 8100 time steps approximately half the system has melted, as indicated by the near-zero values of the order parameter for a number of planes near each of the two free surfaces.

The temperature so obtained from figure above is 1710 ± 30K, in remarkable agreement with the temperature of 1691 ± 20K obtained from the free energy analysis of Broughton and Li.

We conclude from the above evidence that above Tm the grain boundary nucleates the liquid phase which subsequently grows into the crystal, a process requiring thermally-activated diffusion kinetics.

Effect of Free Surfaces

Rather than investigating a single isolated free surface attached to a practically infinite bulk perfect crystal underneath, we consider a thin slab, consisting of 32 (110) planes of atoms, with the outermost planes representing free surfaces. The initial separation between the surfaces (i.e. without atomic relaxations nor thermal expansion) is l1.33a. This film thickness was found to be large enough to ensure that the two surfaces do not interact noticeably over the duration of the simulation 10,000 time steps ≈ 11 ps). In contrast to the simulation of the defect-free perfect crystal and the bicrystal, the border conditions imposed on the simulation cell are now chosen to be periodic only in the two dimensions parallel to the free surfaces, with the latter surrounded by vacuum.

The thin film was heated to 2200K using the same procedure as that used in the grain-boundary studies. Figure, characteristic for all of our thin-film simulations, shows four snapshots of the system starting immediately after the desired simulation temperature of 2200K was reached. The snapshot at t = 0 illustrates the relatively poor coordination of the atoms in the free surface (blue atoms), in contrast to the perfectly coordinated atoms in the interior region of the slab. The order parameter in the left of the figure, well above 0.5 throughout the simulation cell, indicates a well-ordered planar arrangement of the atoms. The smaller values at the two surfaces are primarily due to the larger vibrational amplitudes of the surface atoms.

The following snapshots after 2700, 5400, and 8100 time steps demonstrate the rapid increase in the number of atoms with coordination in excess of four, suggesting that the surface regions have actually melted. As in our grain boundary simulations, we have verified that the atoms are, indeed, in the liquid state by determining the diffusion coefficient in the disordered region. The spreading velocities extracted from these simulations are statistically identical to those obtained from the analysis of the grain boundary simulations.

One might ask whether other extended defects can also act as nucleation centers. This question was investigated in a similar melting study for the face-centered cubic metal copper. In this work we considered the effect of voids of various sizes, with the main conclusion that a void is, indeed, able to nucleate the liquid provided it is large enough and relatively immobile.

From the above evidence, we conclude that above T_m grain boundaries, free surfaces or voids (and we may reasonably expect, lattice dislocations as well) can nucleate the liquid phase which subsequently grows into the crystal-a process requiring thermally-activated diffusion kinetics. Melting is therefore a relatively slow heterogeneous process nucleated at internal or external surfaces. The propagation velocities in figure. typically of the order of a few dozen m/s, are of the same order of magnitude as the velocities obtained from laser-annealing experiments on Si. Given a value, say, of loom/s, obtained at ~500K above the melting point, a single crystal lcm in diameter would require approximately 50µs to melt. This value indicates that melting is by no means an instantaneous process. Also, since our simulation cells contained no intrinsic lattice defects, the above simulations question the validity of theories of melting based on the presence of such defects.

Melting of a Defect-free Crystal

In computer simulations, thermodynamic melting is easily suppressed by elimination of extended defects, for example, via the application of 3-d periodic border conditions to the perfect-crystal simulation cell in figure below. Experimentally, due to the presence of free surfaces and, in most materials, a sufficient atomic concentration of lattice dislocations, superheating is extremely difficult to achieve.

Over half a century ago Born pointed out the existence of an absolute limit to superheating of any crystalline substance. By considering the volume dependence of the normal modes of a crystal lattice, he demonstrated the existence of phonon instability at a certain critical volume of the system. By couching the discussion in terms of the elastic constants (which are known to be intimately connected with the long-wavelength lattice vibrations), Born's phonon instability can be shown to correspond to an elastic instability in the shear constant C_{44} which, for a crystal lattice with cubic symmetry and in the proper coordinate system, is given by $(C_{11}-C_{12})/2$. This instability, occurring when C_{11} and C_{12} become equal, signifies no resistance of the crystal lattice to certain shear strains.

The physical meaning of Born's limit becomes particularly apparent when we consider the Poisson ratio, v, which together with Young's and the shear modulus is needed to characterize the elastic performance of engineering materials. When a tensile strain, $\varepsilon = \Delta 1/1$, is imposed on a tetragonal prism of length 1 and square lateral dimensions d, the consequent contraction, $\Delta d/d$, perpendicular to the direction of the applied stress is governed by the Poisson ratio, defined by:

$$v = (\Delta d / d)/(\Delta 1 / 1)$$

Because the imposed strain is tensile, the resulting volume change of the prism:

$$\Delta V / V = [(d - \Delta d)^2 (1 + \Delta 1) - d^2 1] \sim \varepsilon(1 - 2v)$$

has to be positive, i.e., $0 < v < 0.5$. For a single crystal with cubic symmetry, v may be expressed in terms of C_{11} and C_{12} as follows:

$$v = C_{12}/(C_{11} + C_{12})$$

The condition that $v < 0.5$ therefore requires that C_{12} cannot exceed C_{11}. At the instability point, at which $C_{11} = C_{12}$ (and therefore $v > 0.5$), the volume of the crystal would actually decrease under tension (rather than increase).

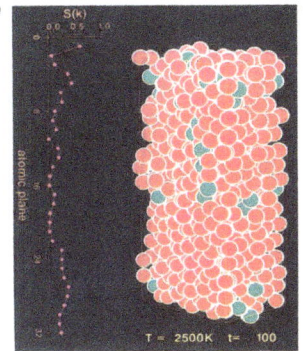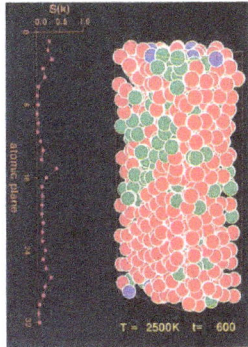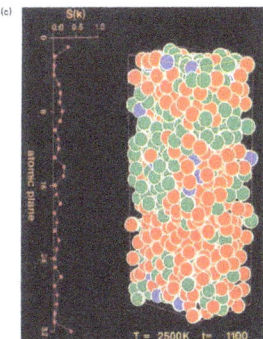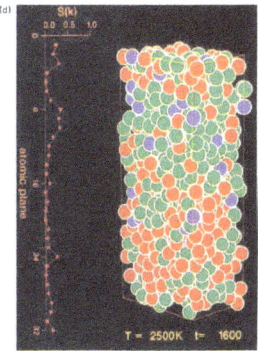

Homogeneous "mechanical melting" of the defectfree perfect crystal of figure above. realized via application of 3-d periodic border conditions. Although the unit-cell dimensions are identical to those of the bicrystal in figure and of the thin slab in figure before this one the 3-d periodic border conditions produce effectively an infinitely large perfect crystal. A comparison with figures demonstrates how rapidly. The process of mechanical melting takes place by comparison with the thermodynamic melting mechanism. (a)-(d) of the figure before this one also demonstrate that the process occurs homogeneously, in contrast to the nucleation-and-growth controlled mechanism of thermodynamic melting.

(a) Only 100 time steps after reaching the simulation temperature, a number of defected, i.e., not perfectly coordinated atoms is found to be scattered randomly throughout the system. Note that the long-range order has practically broken down already throughout the entire crystal, as evidenced by the near-zero values of the related order parameter, $S(k)$, in the left of the figure.

(b)-(d) After 600, 1100, and 1600 time steps, respectively, more and more liquid-like coordinated atoms are found in the simulation cell, and the order parameter fluctuates near zero.

Given that, due to anharmonic effects, virtually all materials expand upon heating, Born's criterion establishes the existence of a maximum volume expansion of a superheated crystal lattice, coupled with a maximum superheating temperature, above which the crystal is mechanically unstable and therefore has to undergo some kind of phase transformation (into the liquid state or into some other crystal structure). The temperature associated with the maximum superheating limit under zero external pressure (thus including the effects of thermal expansion), is referred to as the mechanical melting point, T_s to be distinguished from the thermodynamic melting temperature, T_m. (The subscript "s" refers to "stability".)

In our simulations of metals, the evaluation of C_{11} and C_{12} gives typical values of T, about 20% above T_m. By contrast, we estimate T_s in Si to exceed Tm by as much as 40% (i.e., $T_s \sim 2500K$). In practice it is very difficult, even in simulations, to reach T_s because of statistical fluctuations in the volume and temperature of the system. By gradually stepping up the temperature, we were able to superheat a perfect Si crystal to 2400K, i.e., 700K above Tm; beyond this temperature the crystal could not be stabilized.

Analogous to the free-energy calculation which predicts T_s the determination of T, does not provide information on the mechanism by which the crystalline order breaks down. To investigate this mechanism, we have simulated the melting of a defect-free crystal. (a)-(d) of the above figure illustrate how rapidly the perfect crystal melts above T_s. After a step increase of the simulation temperature from below T_s to 2500 K, only a few hundred MD time steps (i.e., only a few lattice-vibration periods) are required to completely destroy the long-range order within the (110) planes. Moreover, the order-parameter profiles in above figures. (a)-(d) show that planar order is lost simultaneously in all parts of the crystal. This evidence suggests that the liquid phase is formed homogeneously. The above simulation shows clearly the phenomenon of superheating. The reason that it is so easily visible by simulation, in contrast to experiments, is due to the fact that, via the application of 3-d periodic border conditions, surfaces are easily eliminated entirely from the simulation cell.

Characteristics of Thermodynamic and Mechanical Melting

Every crystal, in principle, has two melting points, T_m and T_s. Conceptually the two transitions have distinct physical origins: while thermodynamic melting is governed by the free energies of the liquid and the solid phases, mechanical melting is based on a phonon instability. Since at ambient pressure, the volume expansion required for mechanical melting is always larger than that associated with thermodynamic melting, the free energy always favors thermodynamic over mechanical melting; i.e., $T_s > T_m$. However, as illustrated above, the former requires atomic mobility, and may therefore be kinetically hindered. If a crystal is melted under atmospheric conditions, the thermodynamic state variables usually will be such that high atom mobility in the liquid enables the nucleation and growth of the liquid phase at extended defects. However, if for example by uniformly expanding the crystal, melting is induced at a lower temperature, the consideration of limited atom mobility as a possible hindrance to phase change may be of significant importance. The crystal may, indeed, not be able to disorder at the volume specified by equilibrium thermodynamics until a larger volume is reached where the mechanical instability can occur.

There is considerable experimental evidence that solid-state amorphization, the process in which the long range crystalline order is destroyed by external means (such as mechanical or chemical means, or by irradiation), can proceed by the same two distinct mechanisms as melting and that, in contrast to conventional melting, both types of transition can actually be observed. In a typical melting experiment, the order-disorder transition is induced by increasing the temperature (T) under ambient pressure (P), thus allowing the volume (V) to expand, a procedure guaranteeing high atom mobility at the point (T, P(V)) in thermodynamic phase space where the transition can occur. In a typical solid-state amorphization experiment, by contrast, the temperature is held fixed at some relatively low value, well below Tm ' The role of the irradiation, or of the mechanical or chemical means, in inducing the crystal-to-amorphous transition is to expand the crystal lattice to the coexistence point in phase space where the thermodynamic transition can, in principle, occur. However, relatively low atom mobility gives rise to a competition between the heterogeneous and homogeneous processes, a competition governed by the level of atomic mobility at that point in phase space. Hence, while at higher temperatures mechanical amorphization will be preempted by the thermodynamic type of transition, at lower temperatures this type of transition may be kinetically hindered due to the reduced atom mobility. However, at an even larger volume expansion than that at the thermodynamic coexistence point at a fixed temperature, the ultimate stability limit of the crystal may be reached, thus enabling a fast, homogeneous transition into the liquid state. Due to the low atomic mobility this non-crystalline state appears to be solid, although it merely represents a kinetically arrested liquid.

Let us summarize the three main distinguishing characteristics of thermodynamic and mechanical melting.

a) Whereas thermodynamic melting is based on the free energies of both the crystalline and liquid states, mechanical melting is triggered by phonon instability.

b) Thermodynamic melting is a heterogeneous process, involving nucleation and growth of the liquid phase at extended defects, whereas mechanical melting takes place homogeneously, without the need for the presence of lattice defects.

c) The growth of the liquid phase into the crystal (by propagation of solid-liquid interfaces) requires thermally activated diffusion kinetics in the liquid. Mechanical melting, by contrast, takes place typically within a few lattice vibration periods independent of temperature.

In concluding we point out that, in our view, the present study provides an illustration of several unique features of atomistic simulation, namely the abilities to prescribe precisely the initial system configuration, to control the dynamic environment during the simulations, and to follow the system response in complete detail, all at the atomic level. It seems clear that in future simulation studies of complex physical phenomena these capabilities will be explored to an even greater extent. In doing so one should keep in mind that the significance of the simulation result is always limited by the reality of the interatomic potential function used. For this reason one should look for insight from simulation rather than merely numerical results.

Equilibrium Consideration

Equilibrium constants are determined in order to quantify chemical equilibria. When an equilibrium constant K is expressed as a concentration quotient:

$$K = \frac{[S]^{\sigma}[T]^{\tau}\cdots}{[A]^{\alpha}[B]^{\beta}\cdots}$$

It is implied that the activity quotient is constant. For this assumption to be valid, equilibrium constants must be determined in a medium of relatively high ionic strength. Where this is not possible, consideration should be given to possible activity variation.

The equilibrium expression above is a function of the concentrations [A], [B] etc. of the chemical species in equilibrium. The equilibrium constant value can be determined if any one of these concentrations can be measured. The general procedure is that the concentration in question is measured for a series of solutions with known analytical concentrations of the reactants. Typically, a titration is performed with one or more reactants in the titration vessel and one or more reactants in the burette. Knowing the analytical concentrations of reactants initially in the reaction vessel and in the burette, all analytical concentrations can be derived as a function of the volume (or mass) of titrant added.

The equilibrium constants may be derived by best-fitting of the experimental data with a chemical model of the equilibrium system.

Experimental Methods

There are four main experimental methods. For less commonly used methods. In all cases the range can be extended by using the competition method. An example of the application of this method can be found in palladium(II) cyanide.

Potentiometric Measurements

A free concentration [A] or activity {A} of a species A is measured by means of an ion selective electrode such as the glass electrode. If the electrode is calibrated using activity standards it is assumed that the Nernst equation applies in the form:

$$E = E^0 + \frac{RT}{nF}\ln\{A\}$$

Where E^0 is the standard electrode potential. When buffer solutions of known pH are used for calibration the meter reading will be a pH:

$$pH = \frac{nF}{RT}\left(E^0 - E\right)$$

At 298 K, 1 pH unit is approximately equal to 59 mV.

When the electrode is calibrated with solutions of known concentration, by means of a strong acid–strong base titration, for example, a modified Nernst equation is assumed:

$$E = E^0 + s\log_{10}[A]$$

Where s is an empirical slope factor. A solution of known hydrogen ion concentration may be prepared by standardization of a strong acid against borax. Constant-boiling hydrochloric acid may also be used as a primary standard for hydrogen ion concentration.

Range and Limitations

The most widely used electrode is the glass electrode, which is selective for the hydrogen ion. This is suitable for all acid–base equilibria. $\log_{10}\beta$ values between about 2 and 11 can be measured directly by potentiometric titration using a glass electrode. This enormous range of stability constant values (ca. 100 to 10^{11} is possible because of the logarithmic response of the electrode. The limitations arise because the Nernst equation breaks down at very low or very high pH.

When a glass electrode is used to obtain the measurements on which the calculated equilibrium constants depend, the precision of the calculated parameters is limited by secondary effects such as variation of liquid junction potentials in the electrode. In practice it is virtually impossible to obtain a precision for $\log\beta$ better than ±0.001.

Spectrophotometric Measurements

Absorbance

It is assumed that the Beer–Lambert law applies:

$$A = l\sum \varepsilon c$$

where l is the optical path length, ε is a molar absorbance at unit path length and c is a concentration. More than one of the species may contribute to the absorbance. In principle absorbance may be measured at one wavelength only, but in present-day practice it is common to record complete spectra.

Fluorescence (Luminescence) Intensity

It is assumed that the scattered light intensity is a linear function of species' concentrations:

$$I = \sum \varphi c$$

where φ is a proportionality constant.

Range and Limitations

Absorbance and luminescence: An upper limit on $\log_{10} \beta$ of 4 is usually quoted, corresponding to the precision of the measurements, but it also depends on how intense the effect is. Spectra of contributing species should be clearly distinct from each other.

NMR Chemical Shift Measurements

Chemical exchange is assumed to be rapid on the NMR time-scale. An individual chemical shift δ is the mole-fraction-weighted average of the shifts δ of nuclei in contributing species:

$$\bar{\delta} = \frac{\sum x_i \delta_i}{\sum x_i}$$

Example: the pK_a of the hydroxyl group in citric acid has been determined from 13C chemical shift data to be 14.4. Neither potentiometry nor ultraviolet–visible spectroscopy could be used for this determination.

Range and Limitations

Limited precision of chemical shift measurements also puts an upper limit of about 4 on $\log_{10} \beta$. Limited to diamagnetic systems. 1H NMR cannot be used with solutions of compounds in ^1H_2O.

Calorimetric Measurements

Simultaneous measurement of K and ΔH for 1:1 adducts is routinely carried out using isothermal titration calorimetry. Extension to more complex systems is limited by the availability of suitable software.

Range and Limitations

Insufficient evidence is currently available.

Competition Method

The competition method may be used when a stability constant value is too large to be determined by a direct method. It was first used by Schwarzenbach in the determination of the stability constants of complexes of EDTA with metal ions.

For simplicity consider the determination of the stability constant K_{AB} of a binary complex, AB, of a reagent A with another reagent B:

$$K_{AB} = \frac{[AB]}{[A][B]}$$

where the [X] represents the concentration, at equilibrium, of a species X in a solution of given composition.

A ligand C is chosen which forms a weaker complex with A. The stability constant, K_{AC}, is small enough to be determined by a direct method. For example, in the case of EDTA complexes A is a metal ion and C may be a polyamine such as diethylenetriamine:

$$K_{AC} = \frac{[AC]}{[A][C]}$$

The stability constant, K for the competition reaction:

$$AC + B \rightleftharpoons AB + C$$

can be expressed as:

$$K = \frac{[AB][C]}{[AC][B]}$$

It follows that:

$$K_{AB} = K \times K_{AC}$$

where K is the stability constant for the competition reaction. Thus, the value of the stability constant K_{AB} may be derived from the experimentally determined values of K and K_{AC}.

Computational Methods

It is assumed that the collected experimental data comprise a set of data points. At each ith data point, the analytical concentrations of the reactants, $T_A(i)$, $T_B(i)$ etc. are known along with a measured quantity, y_i, that depends on one or more of these analytical concentrations. A general computational procedure has four main components:

1. Definition of a chemical model of the equilibria.
2. Calculation of the concentrations of all the chemical species in each solution.
3. Refinement of the equilibrium constants.
4. Model selection.

The value of the equilibrium constant for the formation of a 1:1 complex, such as a host-guest species, may be calculated with a dedicated spreadsheet application, Bindfit: In this case step 2 can be performed with a non-iterative procedure and the pre-programmed routine Solver can be used for step 3.

Chemical Model

The chemical model consists of a set of chemical species present in solution, both the reactants added to the reaction mixture and the complex species formed from them. Denoting the reactants by A, B..., each *complex species* is specified by the stoichiometric coefficients that relate the particular combination of *reactants* forming them:

$$pA + qB \cdots \rightleftharpoons A_p B_q \cdots \beta_{pq\cdots} = \frac{[A_p B_q \cdots]}{[A]^p [B]^q \cdots}.$$

When using general-purpose computer programs, it is usual to use cumulative association constants, as shown above. Electrical charges are not shown in general expressions such as this and are often omitted from specific expressions, for simplicity of notation. In fact, electrical charges have no bearing on the equilibrium processes other that there being a requirement for overall electrical neutrality in all systems.

With aqueous solutions the concentrations of proton (hydronium ion) and hydroxide ion are constrained by the self-dissociation of water:

$$H_2O \rightleftharpoons H^+ + OH^- : K'_W = \frac{[H^+][OH^-]}{[H_2O]}$$

With dilute solutions the concentration of water is assumed constant, so the equilibrium expression is written in the form of the ionic product of water:

$$K_W = [H^+][OH^-].$$

When both H⁺ and OH⁻ must be considered as reactants, one of them is eliminated from the model by specifying that its concentration be derived from the concentration of the other. Usually the concentration of the hydroxide ion is given by:

$$[OH^-] = \frac{K_W}{[H^+]}$$

In this case the equilibrium constant for the formation of hydroxide has the stoichiometric coefficients −1 in regard to the proton and zero for the other reactants. This has important implications for all protonation equilibria in aqueous solution and for hydrolysis constants in particular.

It is quite usual to omit from the model those species whose concentrations are considered negligible. For example, it is usually assumed then there is no interaction between the reactants and complexes and the electrolyte used to maintain constant ionic strength or the buffer used to maintain constant pH. These assumptions may or may not be justified. Also, it is implicitly assumed that there are no other complex species present. When complexes are wrongly ignored a systematic error is introduced into the calculations.

Speciation Calculations

A speciation calculation is one in which concentrations of all the species in an equilibrium system are calculated, knowing the analytical concentrations, T_A, T_B etc. of the reactants A, B etc. This means solving a set of nonlinear equations of mass-balance:

$$T_A = [A] + \sum_{1,nk} p\beta_{pq...}[A]^p[B]^q \cdots$$

$$T_B = [B] + \sum_{1,nk} q\beta_{pq...}[A]^p[B]^q \cdots$$

for the free concentrations [A], [B] etc. When the pH (or equivalent e.m.f., E).is measured, the free concentration of hydrogen ions, [H], is obtained from the measured value as:

$$[H] = 10^{-pH} \text{ or } [H] = 10^{\frac{E-E^0}{2.303RT/nF}}$$

and only the free concentrations of the other reactants are calculated. The concentrations of the complexes are derived from the free concentrations via the chemical model.

Some authors include the free reactant terms in the sums by declaring *identity* (unit) β constants for which the stoichiometric coefficients are 1 for the reactant concerned and zero for all other reactants. For example, with 2 reagents, the mass-balance equations assume the simpler form:

$$T_A = \sum_{0,nk} p\beta_{pq}[A]^p[B]^q$$

$$T_B = T_B \sum_{0,nk} q\beta_{pq}[A]^p[B]^q$$

$$\beta_{10} = \beta_{01} = 1$$

In this manner, all chemical species, including the free reactants, are treated in the same way, having been formed from the combination of reactants that is specified by the stoichiometric coefficients.

In a titration system the analytical concentrations of the reactants at each titration point are obtained from the initial conditions, the burette concentrations and volumes. The analytical (total) concentration of a reactant R at the ith titration point is given by:

$$T_R = \frac{R_0 + v_i[R]}{v_0 + v_i}.$$

where R_0 is the initial amount of R in the titration vessel, v_0 is the initial volume, [R] is the concentration of R in the burette and v_i is the volume added. The burette concentration of a reactant not present in the burette is taken to be zero.

In general, solving these nonlinear equations presents a formidable challenge because of the huge range over which the free concentrations may vary. At the beginning, values for the free concentrations must be estimated. Then, these values are refined, usually by means of Newton–Raphson iterations. The logarithms of the free concentrations may be refined rather than the free concentrations themselves. Refinement of the logarithms of the free concentrations has the added advantage of automatically imposing a non-negativity constraint on the free concentrations. Once the free reactant concentrations have been calculated, the concentrations of the complexes are derived from them and the equilibrium constants.

Note that the free reactant concentrations can be regarded as implicit parameters in the equilibrium constant refinement process. In that context the values of the free concentrations are constrained by forcing the conditions of mass-balance to apply at all stages of the process.

Equilibrium Constant Refinement

The objective of the refinement process is to find equilibrium constant values that give the best fit to the experimental data. This is usually achieved by minimising an objective function, U, by the method of non-linear least-squares. First the residuals are defined as:

$$r_i = y_i^{obs} - y_i^{calc}$$

Then the most general objective function is given by:

$$U = \sum_i \sum_j r_i W_{ij} r_j$$

The matrix of weights, W, should be, ideally, the inverse of the variance-covariance matrix of the observations. It is rare for this to be known. However, when it is, the expectation value of U is one, which means that the data are fitted *within experimental error*. Most often only the diagonal elements are known, in which case the objective function simplifies to:

$$U = \sum_i W_{ii} r_i^2$$

with $W_{ij} = 0$ when $j \neq i$. Unit weights, $W_{ii} = 1$, are often used but, in that case, the expectation value of U is the root mean square of the experimental errors.

The minimization may be performed using the Gauss–Newton method. Firstly the objective function is linearised by approximating it as a first-order Taylor series expansion about an initial parameter set, p:

$$U = U^0 + \sum_i \frac{\partial U}{\partial p_i} \delta p_i$$

The increments δp_i are added to the corresponding initial parameters such that U is less than U^0. At the minimum the derivatives $\partial U / \partial p_i$, which are simply related to the elements of the Jacobian matrix, J:

$$J_{jk} = \frac{\partial y_j^{calc}}{\partial p_k}$$

where p_k is the kth parameter of the refinement, are equal to zero. One or more equilibrium constants may be parameters of the refinement. However, the measured quantities represented by y are not expressed in terms of the equilibrium constants, but in terms of the species concentrations, which are implicit functions of these parameters. Therefore, the Jacobian elements must be obtained using implicit differentiation.

The parameter increments δp are calculated by solving the normal equations, derived from the conditions that $\partial U / \partial p = 0$ at the minimum:

$$\left(J^T W J\right) dp = J^T W r$$

The increments δp are added iteratively to the parameters:

$$p^{n+1} = p^n + \delta p$$

where n is an iteration number. The species concentrations and y^{calc} values are recalculated at every data point. The iterations are continued until no significant reduction in U is achieved, that is, until a convergence criterion is satisfied. If, however, the updated parameters do not result in a decrease of the objective function, that is, if divergence occurs, the increment calculation must be modified. The simplest modification is to use a fraction, f, of calculated increment, so-called shift-cutting:

$$p^{n+1} = p^n + f\delta p$$

In this case, the direction of the shift vector, δp, is unchanged. With the more powerful Levenberg–Marquardt algorithm, on the other hand, the shift vector is rotated towards the direction of steepest descent, by modifying the normal equations:

$$\left(J^T WJ + \lambda I\right)\delta p = J^T Wr$$

where λ is the Marquardt parameter and I is an identity matrix. Other methods of handling divergence have been proposed.

A particular issue arises with NMR and spectrophotometric data. For the latter, the observed quantity is absorbance, A, and the Beer–Lambert law can be written as:

$$A_\lambda = l \sum (\varepsilon_{pq..})_\lambda c_{pq..}$$

It can be seen that, assuming that the concentrations, c, are known, that absorbance, A, at a given wavelength, λ, and path length l, is a linear function of the molar absorbptivities, ε. With 1 cm path-length, in matrix notation:

$$A = \varepsilon C$$

There are two approaches to the calculation of the unknown molar absorptivities:

(1) The ε values are considered parameters of the minimization and the Jacobian is constructed on that basis. However, the ε values themselves are calculated at each step of the refinement by linear least-squares $\varepsilon = \left(C^T C\right)^{-1} C^T A$ using the refined values of the equilibrium constants to obtain the speciation. The matrix $\left(C^T C\right)^{-1} C^T$ is an example of a pseudo-inverse.

Golub and Pereyra showed how the pseudo-inverse can be differentiated so that parameter increments for both molar absorptivities and equilibrium constants can be calculated by solving the normal equations.

(2) The Beer–Lambert law is written as:

$$\varepsilon_\lambda = A_\lambda^{-1} C$$

The unknown molar absorbances of all "coloured" species are found by using the non-iterative method of linear least-squares, one wavelength at a time. The calculations are performed once every refinement cycle, using the stability constant values obtaining at that refinement cycle to calculate species' concentration values in the matrix C.

Parameter Errors and Correlation

In the region close to the minimum of the objective function, U, the system approximates to a linear least-squares system, for which:

$$p = \left(J^T W J\right)^{-1} J^T W y^{obs}$$

Therefore, the parameter values are (approximately) linear combinations of the observed data values and the errors on the parameters, p, can be obtained by error propagation from the observations, y^{obs}, using the linear formula. Let the variance-covariance matrix for the observations be denoted by Σ^y and that of the parameters by Σ^p. Then:

$$\Sigma^p = \left(J^T W J\right)^{-1} J^T W \Sigma^y W^T J (J^T W J)^{-1}$$

when $W = (\Sigma^y)^{-1}$, this simplifies to:

$$\Sigma^p = \left(J^T W J\right)^{-1}$$

In most cases the errors on the observations are un-correlated, so that Σ^y is diagonal. If so, each weight should be the reciprocal of the variance of the corresponding observation. For example, in a potentiometric titration, the weight at a titration point, k, can be given by:

$$W_k = \frac{1}{\sigma_E^2 + \left(\dfrac{\partial E}{\partial v}\right)_k^2 \sigma_v^2}$$

where σ_E is the error in electrode potential or pH, $(\partial E / \partial v)_k$ is the slope of the titration curve and σ_v is the error on added volume.

When unit weights are used ($W = I$, $p = (J^T J)^{-1} J^T y$) it is implied that the experimental errors are uncorrelated and all equal: $\Sigma^y = \sigma^2 I$, where σ^2 is known as the variance of an observation of unit weight, and I is an identity matrix. In this case σ^2 is approximated by:

$$\sigma^2 = \frac{U}{n_d - n_p}$$

where U is the minimum value of the objective function and n_d and n_p are the number of data and parameters, respectively:

$$\Sigma^p = \frac{U}{n_d - n_p} \left(J^T J\right)^{-1}$$

In all cases, the variance of the parameter p_i is given by Σ^p_{ii} and the covariance between parameters p_i and p_j is given by Σ^p_{ij}. Standard deviation is the square root of variance. These error estimates reflect only random errors in the measurements. The true uncertainty in the parameters is larger due to the presence of systematic errors—which, by definition, cannot be quantified.

Note that even though the observations may be uncorrelated, the parameters are always correlated.

Derived Constants

When cumulative constants have been refined it is often useful to derive stepwise constants from them. The general procedure is to write down the defining expressions for all the constants involved and then to equate concentrations. For example, suppose that one wishes to derive the pKa for removing one proton from a tribasic acid, LH_3, such as citric acid.

$$L^{3-} + H^+ \rightleftharpoons LH^{2-} : [LH^{2-}] = \beta_{11}[L^{3-}][H^+]$$
$$L^{3-} + 2H^+ \rightleftharpoons LH_2^- : [LH_2^-] = \beta_{12}[L^{3-}][H^+]^2$$
$$L^{3-} + 3H^+ \rightleftharpoons LH_3 : [LH_3] = \beta_{13}[L^{3-}][H^+]^3$$

The stepwise *association* constant for formation of LH_3 is given by:

$$LH2^- + H^+ \rightleftharpoons LH3; \quad [LH3] = K[LH2^-][H^+]$$

Substitute the expressions for the concentrations of LH_3 and LH_2^- into this equation:

$$\beta_{13}[L^{3-}][H^+]^3 = K\beta_{12}[L^{3-}][H^+]^2[H^+]$$

whence,

$$\beta_{13} = K\beta_{12}; \; K = \frac{\beta_{13}}{\beta_{12}}$$

and since $pK_a = -\log_{10} 1/k$ its value is given by:

$$pK_{a_1} = \log_{10} \beta_{13} - \log_{10} \beta_{12}$$
$$pK_{a_2} = \log_{10} \beta_{12} - \log_{10} \beta_{11}$$
$$pK_{a_3} = \log_{10} \beta_{11}$$

Note the reverse numbering for pK and log β. When calculating the error on the stepwise constant, the fact that the cumulative constants are correlated must accounted for. By error propagation:

$$\sigma_K^2 = \sigma_{\beta_{12}}^2 + \sigma_{\beta_{13}}^2 - 2\sigma_{\beta_{12}}\sigma_{\beta_{13}}\rho_{12,13}$$

and

$$\sigma_{\log_{10} K} = \frac{\sigma_K}{K}.$$

Model Selection

Once a refinement has been completed the results should be checked to verify that the chosen model is acceptable. Generally speaking, a model is acceptable when the data are fitted within experimental error, but there is no single criterion to use to make the judgement. The following should be considered.

Objective Function

When the weights have been correctly derived from estimates of experimental error, the expectation value of $\dfrac{U}{nd - np}$ is 1. It is therefore very useful to estimate experimental errors and derive some reasonable weights from them as this is an absolute indicator of the goodness of fit.

When unit weights are used, it is implied that all observations have the same variance. $\dfrac{U}{nd - np}$ is expected to be equal to that variance.

Parameter Errors

One would want the errors on the stability constants to be roughly commensurate with experimental error. For example, with pH titration data, if pH is measured to 2 decimal places, the errors of $\log_{10} \beta$ should not be much larger than 0.01. In exploratory work where the nature of the species present is not known in advance, several different chemical models may be tested and compared. There will be models where the uncertainties in the best estimate of an equilibrium constant may be somewhat or even significantly larger than σ_{pH}, especially with those constants governing the formation of comparatively minor species, but the decision as to how large is acceptable remains subjective. The decision process as to whether or not to include comparatively uncertain equilibria in a model, and for the comparison of competing models in general can be made objective and has been outlined by Hamilton.

Distribution of Residuals

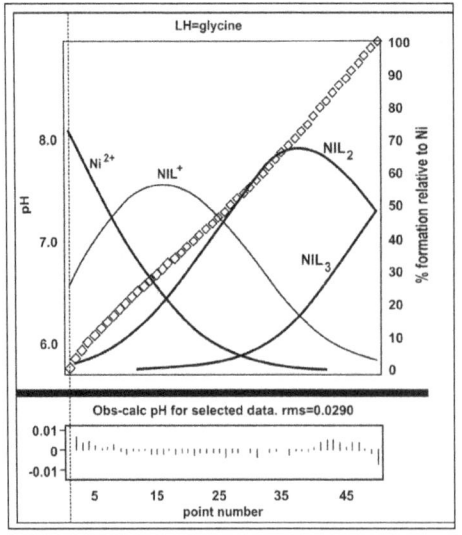

At the minimum in U the system can be approximated to a linear one, the residuals in the case of unit weights are related to the observations by:

$$r = y^{obs} - J\left(J^{T}T\right)^{-1} J^{T} y^{obs}$$

The symmetric, idempotent matrix $J(J^TT)^{-1}J$ is known in the statistics literature as the hat matrix, H. Thus,

$$r = (I-H)y^{obs}$$

and

$$M^r = (I-H)M^y(I-H)$$

where I is an identity matrix and M^r and M^y are the variance-covariance matrices of the residuals and observations, respectively. This shows that even though the observations may be uncorrelated, the residuals are always correlated.

The diagram at the right shows the result of a refinement of the stability constants of $Ni(Gly)^+$, $Ni(Gly)_2$ and $Ni(Gly)_3^-$ (where GlyH = glycine). The observed values are shown a blue diamonds and the species concentrations, as a percentage of the total nickel, are superimposed. The residuals are shown in the lower box. The residuals are not distributed as randomly as would be expected. This is due the variation of liquid junction potentials and other effects at the glass/liquid interfaces. Those effects are very slow compared to the rate at which equilibrium is established.

Physical Constraints

Some physical constraints are usually incorporated in the calculations. For example, all the concentrations of free reactants and species must have positive values and association constants must have positive values.

With spectrophotometric data the molar absorptivity (or emissivity) values should all be positive. Most computer programs do not impose this constraint on the calculations.

Other Models

If the model is not acceptable, a variety of other models should be examined to find one that best fits the experimental data, within experimental error. The main difficulty is with the so-called minor species. These are species whose concentration is so low that the effect on the measured quantity is at or below the level of error in the experimental measurement. The constant for a minor species may prove impossible to determine if there is no means to increase the concentration of the species.

Crystalline Melting Temperature

The temperature at which polymer changes from a viscous liquid to a micro-crystalline solid is the crystalline melting point of the polymer.

- Tm is not as sharp as melting point of other solids but is analogous to it.

- Accompanied by changes in density, refractive index, heat capacity,etc.

- Tg=0.5Tm Or Tg=(2/3)Tm (for most of the polymers).

Kinetics of Polymer Crystallization

Under certain conditions, polymers cooled from the melt can arrange into regular crystalline structures. The formation of these polymer crystals involves two processes that occur spontaneously when a polymer melt is cooled below its melting point. The first process is the formation of stable primary nuclei which itself are the result of fluctuations in density or order of supercooled melt. The growth rate of these nuclei requires sufficient mobility which increases with increasing temperature, whereas the rate of nucleation increases with decreasing temperature. Thus, the overall process displays a maximum growth rate somewhere between the melting point and glass transition temperature. Due to the larger rate of nucleation and smaller growth rate, the size of the crystallites (spherulites) decreases at lower temperatures whereas their number per unit volume increases. The lower mobility of the polymer chains at low temperatures also leads to less perfect crystallites (more defects in each lamellae).

Crystallizaton does not always occur, that is, some polymers are 100 percent amorphous. This is the case when the growth rate of the polymer crystallites is much slower than the cooling rate. Or in other words, crystallization only proceeds if the temperature stays in the appropriate temperature range for an extended time long enough to convert the melt into crystals. However, some portion of the polymer will always remain in a non-crystalline (irregular or amorphous) state. It's fraction will depend on factors such as cooling rate, molecular weight, and degree of branching.

In some cases, spontaneous nucleation is not sufficient to reach acceptable growth rates. In that case, small particles are added to the melt that act as seeds or nucleation agents. Partial melting can also improve the degree of crystallinity because the remaining crystals in the melt act as nuclei.

The growth rate as a function of temperature has more or less the shape of a bell curve with a maximum growth rate in between the glass transition and equilibrium melting temperature; starting from the melting point, as the temperature is lowered, the (spherulite) growth rate increases until it reaches a maximum at an intermediate temperature. Further decrease in temperature leads to a decrease in crystal growth rate and below the glass transition temperature it becomes virtually zero because molecular motion is too slow to allow for crystallization to take place.

The morphology of the crystals and and the degree of crystallinity depend on the type and structure of the polymer, and on the growth conditions. A narrow molecular weight distribution and a high molecular weight tend to increase the crystallinity. The same is true for strong intermolecular forces and stiff chain backbones because the molecules prefer an ordered arrangement with maximum packing density to maximize the number of secondary bonds. Bulky side groups and branching, on the other hand, lower the crystallinity because with increasing size of the side groups it becomes progressively more difficult for the polymer to fold and align itself along the crystal growth direction. The degree of crystallinity also depends on the tacticity of the polymer. The greater the order in a macromolecule the greater the likelihood of the molecule to undergo crystallization. In fact, most atactic polymers do not crystallize.

Generalized Avrami Equation

Isothermal crystallization is often described by the Avrami equation:

$$\theta_c(t) = 1 - \exp[-k_n(T)\, t^n]$$

Where $\theta_c(t)$ is the fraction of transformed material at time t, n is the Avrami exponent, and k the overall crystallization rate constant. The later parameter combines the contribution of both nucleation and growth.

The Avrami index in the equation above can be split into two indices:

$$n = n_d + n_n$$

where n_d represents the dimensionality of the growing crystals. This parameter can only have the values 1,2, or 3. In the case of polymers, only 2 or 3 are encountered. $n_d = 2$ describes two dimensional lamellar growth and $n_d = 3$ three dimensional radial growth, i.e. spherulites. n_n describes the time dependence of nucleation. This parameter has values between 0 and 1, where 0 corresponds to instantaneous nucleation an 1 to sporadic nucleation. A value of about 0.5 is often indicative for diffusion controlled growth, because diffusion is proportional to square root of time. Thus, the overall Avrami index, n, typically varies between 2 and 4.

In practice, the Avrami equation is normally written in its logarithmic form:

$$\ln\{-\ln[1 - \theta_c(t)]\} = \ln Q = \ln k_n + n \ln t$$

A plot of $\ln Q$ versus $\ln t$ yields a straight line from which the Avrami parameters k_n (antilog of intercept) and n (slope) can be readily extracted.

Crystallization Kinetic According to Avrami Model

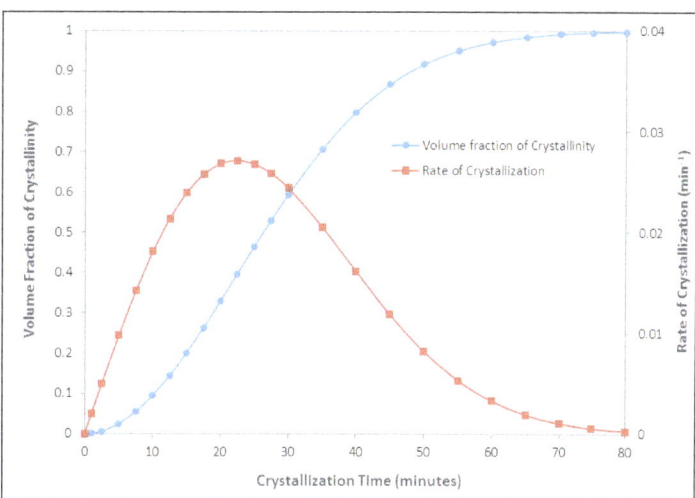

The figure above shows the normalized volume fraction of crystallinity and the rate of crystallization as a function of time for two dimensional growth kinetic ($n = 2$) and constant crystallization rate ($k = 10^{-3}$ min.$^{-2}$). The volume fraction has been normalized with the maximum possible crystallinity at infinite time and the rate of crystallization is the derivative of the Avrami equation with respect to time:

$$d\theta_c(t) / dt = n\, t^{n-1}\, k_n(T) \exp[-k_n(T)\, t^n]$$

The crystallization half-time of this hypothetical crystal growth process is,

$$t_{1/2} = [\ln 2 / k_n(T)]^{1/n} = [\ln 2 / 10^{-3}]^{1/2} = 26.3 \text{ min},$$

and the maximum rate of crystallization is,

$$t_{max} = [(n-1) / nk_n(T)]^{1/n} = [1 / 2 \cdot 10^{-3}]^{1/2} = 22.4 \text{ min}.$$

Glass Transition

The glass transition is a property of only the amorphous portion of a semi-crystalline solid. The crystalline portion remains crystalline during the glass transition.

At a low temperature the amorphous regions of a polymer are in the glassy state. In this state the molecules are frozen on place. They may be able to vibrate slightly, but do not have any segmental motion in which portions of the molecule wiggle around. In the glassy state, the motion of the red molecule in the schematic diagram at the right would not occur. When the amorphous regions of a polymer are in the glassy state, it generally will be hard, rigid, and brittle.

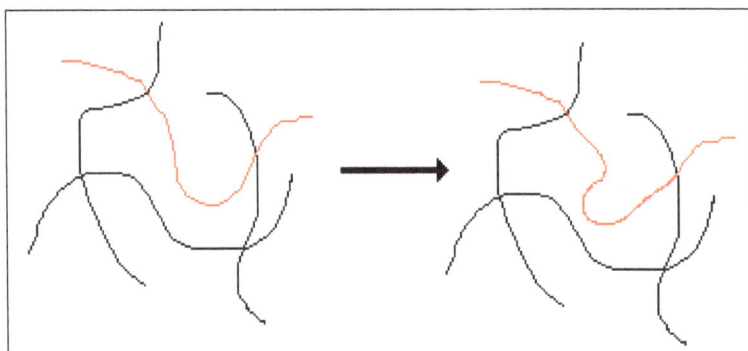

If the polymer is heated it eventually will reach its glass transition temperature. At this temperature portions of the molecules can start to wiggle around as is illustrated by the red molecule in the diagram above. The polymer now is in its rubbery state. The rubbery state lends softness and flexibility to a polymer.

You may have experienced the glass transition of chewing gum. At body temperature the gum is soft and pliable, which is characteristic of an amorphous solid in the rubbery state. If you put a cold drink in your mouth or hold an ice cube on the gum, it becomes hard and rigid. The glass transition temperature of the gum is somewhere between 0 °C and 37 °C.

Comparison with Melting

The glass transition is not the same as melting.

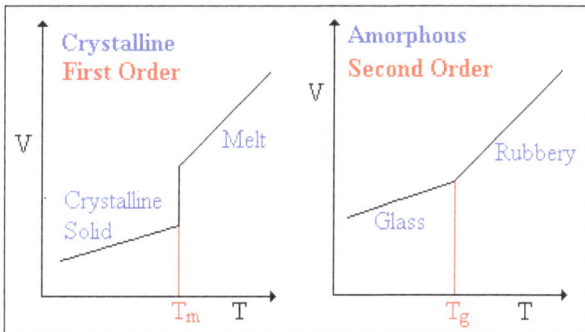

Thermodynamic transitions are classified as being first-or second-order. In a first-order transition there is a transfer of heat between system and surroundings and the system undergoes an abrupt volume change. In a second-order transition, there is no transfer of heat, but the heat capacity does change. The volume changes to accomodate the increased motion of the wiggling chains, but it does not change discontinuously.

Illustrative plots of specific volume vs. temperature are shown at the right for amorphous and crystalline polymers.

Glass Transition Temperature

When an amorphous polymer is heated, the temperature at which it changes from a glass to the rubbery form is called the glass transition temperature, T_g. A given polymer sample does not have a unique value of T_g because the glass phase is not at equilibrium. The measured value of T_g will depend on the molecular weight of the polymer, on its thermal history and age, on the measurement method, and on the rate of heating or cooling. Approximate glass transition temperatures of a few polymers are shown below.

Polymer		T_g (°C)
Polyethylene	(LDPE)	-125
Polypropylene	(atactic)	-20
Poly(vinyl acetate)	(PVAc)	28
Poly(ethyleneterephthalate)	(PET)	69
Poly(vinyl alcohol)	(PVA)	85
Poly(vinyl chloride)	(PVC)	81
Polypropylene	(isotactic)	100
Polystyrene		100
Poly(methylmethacrylate)	(atactic)	105

References

- Crystal, science: britannica.com, Retrieved 10 February, 2019

- Definition-of-crystallize-605854: thoughtco.com, Retrieved 18 April, 2019

- What-is-crystallization-and-what-are-the-methods-of-crystallization: syrris.com, Retrieved 4 January, 2019

- Martell, A. E.; Motekaitis, R. J. (1992). The Determination and Use of Stability Constants. Wiley-VCH. ISBN 0471188174

- Crystallization-Kinetics, polymer-physics: polymerdatabase.com, Retrieved 2 May, 2019

- GlassTrans, polymer-resources: uscupstate.edu, Retrieved 15 March, 2019

6

Thermodynamics of Polymer Solutions and Rheological Properties

The branch of science which focuses on the study of heat and temperature in polymer solutions is known as thermodynamics. The scientific discipline which studies the deformation of polymeric fluids under external stress is known as polymer rheology. The diverse aspects of thermodynamics of polymer solutions as well as the rheological properties of polymers have been thoroughly discussed in this chapter.

Thermodynamics

Thermodynamics is the science of the relationship between heat, work, temperature, and energy. In broad terms, thermodynamics deals with the transfer of energy from one place to another and from one form to another. The key concept is that heat is a form of energy corresponding to a definite amount of mechanical work.

Heat was not formally recognized as a form of energy until about 1798, when Count Rumford (Sir Benjamin Thompson), a British military engineer, noticed that limitless amounts of heat could be generated in the boring of cannon barrels and that the amount of heat generated is proportional to the work done in turning a blunt boring tool. Rumford's observation of the proportionality between heat generated and work done lies at the foundation of thermodynamics. Another pioneer was the French military engineer Sadi Carnot, who introduced the concept of the heat-engine cycle and the principle of reversibility in 1824. Carnot's work concerned the limitations on the maximum amount of work that can be obtained from a steam engine operating with a high-temperature heat transfer as its driving force. Later that century, these ideas were developed by Rudolf Clausius, a German mathematician and physicist, into the first and second laws of thermodynamics, respectively.

The most important laws of thermodynamics are:

- The zeroth law of thermodynamics. When two systems are each in thermal equilibrium with a third system, the first two systems are in thermal equilibrium with each other. This property makes it meaningful to use thermometers as the "third system" and to define a temperature scale.

- The first law of thermodynamics or the law of conservation of energy. The change in a system's internal energy is equal to the difference between heat added to the system from its surroundings and work done by the system on its surroundings.

- The second law of thermodynamics. Heat does not flow spontaneously from a colder region to a hotter region, or, equivalently, heat at a given temperature cannot be converted entirely into work. Consequently, the entropy of a closed system, or heat energy per unit temperature, increases over time toward some maximum value. Thus, all closed systems tend toward an equilibrium state in which entropy is at a maximum and no energy is available to do useful work. This asymmetry between forward and backward processes gives rise to what is known as the "arrow of time."

- The third law of thermodynamics. The entropy of a perfect crystal of an element in its most stable form tends to zero as the temperature approaches absolute zero. This allows an absolute scale for entropy to be established that, from a statistical point of view, determines the degree of randomness or disorder in a system.

Although thermodynamics developed rapidly during the 19th century in response to the need to optimize the performance of steam engines, the sweeping generality of the laws of thermodynamics makes them applicable to all physical and biological systems. In particular, the laws of thermodynamics give a complete description of all changes in the energy state of any system and its ability to perform useful work on its surroundings.

Free Energy of Mixing

A solution is created when two or more components mix homogeneously to form a single phase. Studying solutions is important because most chemical and biological life processes occur in systems with multiple components. Understanding the thermodynamic behavior of mixtures is integral to the study of any system involving either ideal or non-ideal solutions because it provides valuable information on the molecular properties of the system.

Most real gases behave like ideal gases at standard temperature and pressure. This allows us to combine our knowledge of ideal systems and solutions with standard state thermodynamics in order to derive a set of equations that quantitatively describe the effect that mixing has on a given gas-phase solution's thermodynamic quantities.

Gibbs Free Energy of Mixing

Unlike the extensive properties of a one-component system, which rely only on the amount of the system present, the extensive properties of a solution depend on its temperature, pressure and composition. This means that a mixture must be described in terms of the partial molar quantities of its components. The total Gibbs free energy of a two-component solution is given by the expression:

$$G = n_1 \overline{G}_1 + n_2 \overline{G}_2$$

where:

- G is the total Gibbs energy of the system,

- n_1 is the number of moles of component i, and

- $\overline{G}_1$ is the partial molar Gibbs energy of component i.

The molar Gibbs energy of an ideal gas can be found using the equation:

$$\bar{G} = \bar{G}^{\circ} + RT \ln \frac{P}{1bar}$$

where, $\bar{G}^{\circ}$ is the standard molar Gibbs energy of the gas at 1 bar, and P is the pressure of the system. In a mixture of ideal gases, we find that the system's partial molar Gibbs energy is equivalent to its chemical potential, or that:

$$\bar{G}_1 = \mu_1$$

This means that for a solution of ideal gases, Equation $\bar{G} = \bar{G}^{\circ} + RT \ln \frac{P}{1bar}$ can become:

$$\bar{G}_1 = \mu_1 = \mu_1^{\circ} + RT \ln \frac{P_1}{1bar} \mu$$

where:

- μ_i is the chemical potential of the ith component,

- μ_i° is the standard chemical potential of component i at 1 bar, and

- P_i is the partial pressure of component i.

Now pretend we have two gases at the same temperature and pressure, gas 1 and gas 2. The Gibbs energy of the system before the gases are mixed is given by Equation $G = n_1\bar{G}_1 + n_2\bar{G}_2$, which can be combined with Equation $\bar{G}_1 = \mu_1 = \mu_1^{\circ} + RT \ln \frac{P_1}{1bar}$ to give the expression

$$G_{initial} = n_1(\mu_1^{\circ} + RT \ln P) + n_2(\mu_2^{\circ} + RT \ln P).$$

If gas 1 and gas 2 are then mixed together, they will each exert a partial pressure on the total system, P_1 and P_2, so that $P_1 + P_2 = P$. This means that the final Gibbs energy of the final solution can be found using the equation:

$$G_{final} = n_1(\mu_1^{\circ} + RT \ln P_1) + n_2(\mu_2^{\circ} + RT \ln P_2)$$

The Gibbs energy of mixing, $\Delta_{mix}G$, can then be found by subtracting $G_{initial}$ from G_{final}.

$$\Delta_{mix}G = Gfinal - G_{initial}$$

$$= n_1 RT \ln \frac{P_1}{P} + n_2 RT \ln \frac{P_2}{P}$$

$$= n_1 RT \ln x_1 + n_2 RT \ln x_2$$

where,

$$P_i = x_i P$$

and X_i is the mole fraction of gas i. This equation can be simplified further by knowing that the mole fraction of a component is equal to the number of moles of that component over the total moles of the system, or:

$$X_i = \frac{n_i}{n}.$$

Equation $= n_1 RT \ln X_1 + n_2 RT \ln X_2$ then becomes:

$$\Delta_{mix} G = nRT(x_1 \ln x_1 + x_2 \ln x_2)$$

This expression gives us the effect that mixing has on the Gibbs free energy of a solution. Since x_1 and x_2 are mole fractions that range from 0 to 1, we can conclude that $\Delta_{mix} G$ will be a negative number. This is consistent with the idea that gases mix spontaneously at constant pressure and temperature.

Entropy of Mixing

Figure shows that when two gases mix, it can really be seen as two gases expanding into twice their original volume. This greatly increases the number of available microstates, and so we would therefore expect the entropy of the system to increase as well.

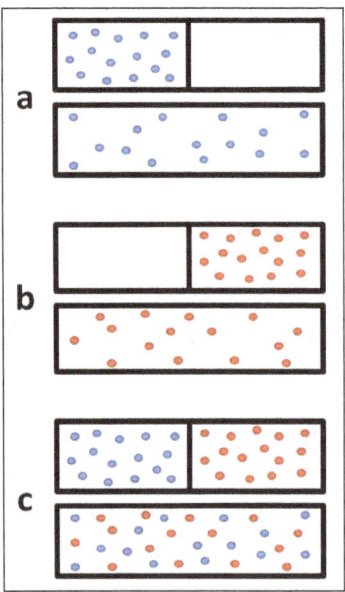

The mixing of two gases can be seen as two expansions. (a) Expansion of gas 1 alone when teh barrier is removed. The molecules have twice as many microstates in the open box. (b) Expansion of gas 2 along. (c) the simultaneous expansion of gases 1 and 2 is equivalent to mixing.

Thermodynamic studies of an ideal gas's dependence of Gibbs free energy of temperature have shown that:

$$\left(\frac{dG}{dT}\right) p = -S.$$

This means that differentiating Equation $\Delta_{mix}G = nRT(x_1 \ln x_1 + x_2 \ln x_2)$ at constant pressure with respect to temperature will give an expression for the effect that mixing has on the entropy of a solution. We see that:

$$\left(\frac{dG_{mix}}{dT}\right)_p = nR(x_1 \ln x_1 + x_2 \ln x_2)$$

$$= -\Delta_{mix}S$$

$$\Delta_{mix}S = -nR(x_1 \ln x_1 + x_2 \ln x_2)$$

Since the mole fractions again lead to negative values for $\ln x_1$ and $\ln x_2$, the negative sign in front of the equation makes $\Delta_{mix}S$ positive, as expected. This agrees with the idea that mixing is a spontaneous process.

Enthalpy of Mixing

We know that in an ideal system $\Delta G = \Delta H - T\Delta S$, but this equation can also be applied to the thermodynamics of mixing and solved for the enthalpy of mixing so that it reads:

$$\Delta_{mix}H = \Delta_{mix}G + T\Delta_{mix}S$$

Plugging in our expressions for $\Delta_{mix}G$ (Equation $\Delta_{mix}G = nRT(x_1 \ln x_1 + x_2 \ln x_2)$) and $\Delta_{mix}S$ (Equation $\Delta_{mix}S = -nR(x_1 \ln x_1 + x_2 \ln x_2)$), we get:

$$\Delta_{mix}H = nRT(x_1 \ln x_1 + x_2 \ln x_2) + T[-nR(x_1 \ln x_1 + x_2 \ln x_2)] = 0.$$

This result makes sense when considering the system. The molecules of ideal gas are spread out enough that they do not interact with one another when mixed, which implies that no heat is absorbed or produced and results in a $\Delta_{mix}H$ of zero. Figure illustrates how $T\Delta_{mix}S$ and $\Delta_{mix}G$ change as a function of the mole fraction so that $\Delta_{mix}H$ of a solution will always be equal to zero (this is for the mixing of two ideal gasses).

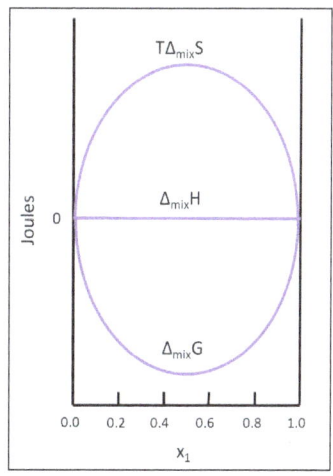

A graph of T∆mixS, T∆mixH, and T∆mixG as a function of x1 for the mixing of two ideal gases.

Phase Behaviour of Polymer Solution

Mixing, Alloying, and Blending are familiar terms that describe the process of combining two or more elements, compounds, or molecular species into a single product. The associated mechanical, chemical, electrical, and various other properties are largely determined by the resulting phase behavior. For example, copper and zinc form a single solid phase known as brass that is mechanically superior to either constituent alone. Nonequilibrium microstructured phases in steel, produced by adding carbon and other elements to iron, can impart exceptional strength and hardness. Mixtures of oil and water that normally macroscopically phase separate can be finely dispersed by the addition of small amounts of surfactant, which can lead to gross changes in wetting and flow properties. All of these effects can be obtained with polymer-polymer mixtures, sometimes with the use of a single pair of monomeric building blocks. Such a diverse range of phase behaviors and the accompanying breadth of materials applications stem from the unparalleled range of molecular architectures that can be realized with polymers.

Molecular Architecture

The number of molecular configurations available to a pair of chemically distinct polymer species (here we define a polymer as a sequence of many repeat units) is almost unlimited. A small subset of the possibilities, representing four commonly encountered types of combinations, is illustrated in the figure below.

Typical polymer-polymer (dashed and solid curves) molecular architectures
attainable through the polymerization of two distinct monomers.

The simplest and best understood among these is the case of binary homopolymer mixtures. At equilibrium such mixtures consist of either one or two phases (neglecting crystallization). In the event of phase separation, interfacial tension favors a reduction in surface area that leads to macroscopic segregation as sketched in the figure below. However, polymer melts (materials that have been heated above the glass transition temperature Tg) are extremely viscous so that phase-separated homopolymers essentially never reach an equilibrium morphology. Consequently, molecular architecture, which strongly influences polymer mobility, plays an important role in the evolution of phase morphology. Branching in particular disrupts the basic mechanism of polymer motion (known as reptation) and leads to significant increases in viscosity. Since the majority of experimental and theoretical studies of homopolymer mixtures has been directed at linear macromolecules, the discussion is limited to this type of mixture. Nevertheless, the qualitative features identified are applicable to star and branched architectures provided the governing thermodynamic and dynamic relations are appropriately modified.

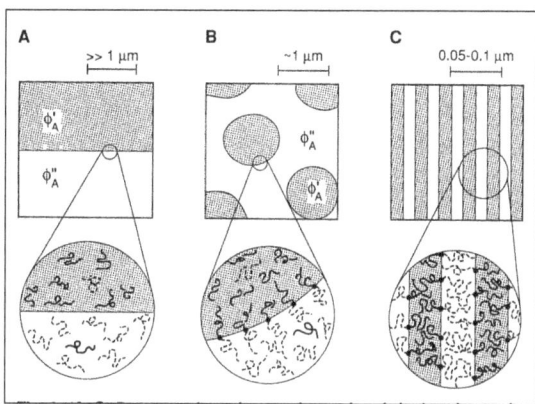

The above figure shows (A-C) Representative polymer-polymer phase behaviors that can be realized with different molecular architectures. Macrophase separation (A) results when thermodynamically incompatible linear homopolymers are mixed. The covalent bond between blocks in a diblock copolymer leads to microphase segregation (C). A mixed architecture of linear homo polymers and the corresponding diblock copolymer produces a surfactant-like stabilized intermediate-scale phase separation (B).

Linear and star homopolymers with well-defined molecular weights and narrow molecular weight distributions are readily synthesized by using anionic polymerization techniques. Alternative synthetic procedures, such as those used in the commercial production of linear and branched homopolymers, usually lead to less-ideal molecular structures that are not as attractive for model studies of polymer-polymer phase behavior. Formation of block copolymers is an alternative method of mixing chemically different polymers. This seemingly minor chemical modification of covalently bonding polymer A to polymer B can be performed in inumerable ways, each with particular thermodynamic and dynamic consequences. Three classes of commonly encountered block copolymer architectures are identified in the above figure.

Star block copolymers represent the best defined group of block copolymers available at present. Model materials with specific arm number, composition, and molecular weight, and with a narrow molecular weight distribution can be prepared by anionic polymerization techniques. These attributes have led many experimentalists and nearly all theoreticians to focus on this class of block copolymers.

Graft and multiblock copolymers compensate with commercial importance what is lacking in architectural perfection. Synthetic routes to these materials generally lead to broad molecular weight distributions and variable graft and block functionalities. Graft block copolymers are frequently blended with homopolymers, whereas multiblock copolymers, such as polyurethanes, are usually prepared neat.

In all cases the most significant factor in determining blockcopolymer phase behavior is the covalent bond that restricts macroscopic separation of chemically dissimilar polymer blocks. This constraint leads to the formation of microscopic heterogeneities in composition at length scales comparable to the molecular dimensions, that is, around 50 to 1000 A, which contrasts sharply with the macroscopic phase separation associated with binary mixtures. An example of such a microstructured (and ordered) phase is given in figure for a symmetric (50% by volume component A) diblock copolymer. Such microphase segregation characterizes all block copolymers. However, detailed theoretical predictions for the associated phase behavior (such as the shape and packing symmetry of the microstructures) are only available for certain model star block copolymers.

Thus far specific molecular architectures have been considered individually. Many multicomponent polymer applications exploit mixed molecular architectures, such as combinations of graft block copolymers and homopolymer mixtures. In this case block copolymers can function as surfactants, localizing at interfaces in a two phase system and thereby reduce the equilibrium phase size, as illustrated in figure. Molecular architecture plays a significant role in delimiting the molecular weight, composition, and concentration of block copolymer required to obtain a particular particle size. Homopolymer architecture, molecular weight, and composition also contribute to the ultimate phase behavior. For example, if the molecular weight of homopolymer is significantly greater than that of diblock copolymer, a three-phase system (A-rich, B-rich, and A-B-rich) can result rather than the surface-stabilized two-phase state depicted in figure. Understanding of mixed architecture systems is rather limited, although recent progress can be found for diblock copolymer-linear homopolymer mixtures. Notable examples of commercially important mixtures of graft block copolymers and linear homopolymers are high-impact polystyrene (HIPS) and acrylonitrile-butadiene-styrene (ABS) resins. Both materials derive superior impact strength from stabilized two-phase morphologies similar to that shown in the above figure.

In the following sections macrophase separation and microphase segregation is examined in terms of the simplest and most studied molecular architecture in each category, binary linear homopolymer mixtures and diblock copolymers, respectively. Although the understanding of these model systems is rather well developed, it should be recognized at the outset that the broader scope summarized by the figure (including mixed architectures) remains largely unexplored at a fundamental level.

Thermodynamic Parameters

Equilibrium polymer-polymer phase behavior is controlled by four factors: (i) molecular architecture, (ii) choice of monomers, (iii) composition, and (iv) degree of polymerization. Composition generally refers to the overall volume fraction of a component. For homopolymer mixtures this parameter is frequently associated with the symbol φ while recent custom assigns f to the star block-copolymer composition. Experimentally, φ is varied by simply changing the mixture stoichiometry. Shifting f requires the synthesis of a new material, since individual blocks are chemically bonded together. The degree of polymerization is the number of repeat units (monomers) that make up a polymer chain. Most thermodynamic theories presume a single repeat unit volume, although in practice chemically different repeat units rarely occupy equal volumes. Therefore, it is convenient to define a segment volume V corresponding to either of the repeat unit volumes (V_A or V_B), or any suitable mean repeat unit volume, for example, $V_A^{1/2} V_B^{1/2}$. With this definition the number of segments per polymer molecule is $N = pVX/M$, where p and M are the polymer density and molecular weight, and X is Avogadro's number. Note that based on this convention $f = N_A/N$.

The choice of a particular pair of monomers establishes the sign and magnitude of the energy of mixing, which can be approximated by the Flory-Huggins segment-segment interaction parameter X,

$$X = \frac{1}{k_B T}\left[\in_{AB} - \frac{1}{2}(\in_{AA} + \in_{BB})\right]$$

Where, $\in ij$ represents the contact energy between i and j segments, and k_B is the Boltzmann constant. A negative value of X results from a favorable energy of mixing, that is, A-B segment-segment contacts on average produce a lower system energy than the sum of A-A and B-B contacts. Certain types of specific A-B interactions, such as hydrogen bonding, can result in a negative X parameter. Positive values of X occur when the net system energy increases upon forming A-B contact pairs from unmixed (pure) components. Most nonpolar polymers, such as polyethylene, polystyrene, and polyisoprene, are characterized by dispersive (van der Waals) interactions that can be represented by,

$$\in_{ij} = -\sum_{i,j}\frac{3}{4}\frac{Ii\,Ij}{Ii+Ij}\frac{ai\,aj}{r_{ij}^6}$$

where, r_{ij} is the segment-segment separation, ex and I are the segment polarizability and ionization potential, respectively. If there is no volume change $(\Delta V_m = 0)$ or preferential segment orientation upon mixing, Equations $= \frac{1}{k\,T}\left[\in_{AB} - \frac{1}{-}(\in_{AA} + \in_{BB})\right]$ and $\in_{ij} = -\sum_{i,j}\frac{3}{4}\frac{Ii\,Ij}{Ii+Ij}\frac{ai\,aj}{r_{ij}^6}$ can be rearranged to,

$$X = \frac{3}{16}\frac{I}{k_B T}\frac{z}{V^2}(a_A - a_B)^2$$

where a cubic lattice is assumed with $I_j = I_j = I$ (valid within 10% for most hydrocarbon polymers) arrd all but the z nearest-neighbor contacts are neglected. Thus, for polymer mixtures governed solely by dispersion interactions $X \geq 0$.

In practice the assumptions made in deriving equation $X = \frac{3}{16}\frac{I}{k_B T}\frac{z}{V^2}(\alpha_A - \alpha_B)^2$ are rarely realized. Significant deviations from incompressibility $(\Delta V_m \neq 0) \cong$ necessitate equation -of-state corrections to X. Anisotropic monomer structures may lead to nonrandom segment packing that must be absorbed in X as an excess entropy of mixing. These effects are usually accounted for by assuming that:

$$X = \alpha T^{-1} + \beta$$

where, α and β represent experimentally determined enthalpy and excess entropy coefficients for a particular composition. In general ex and β may depend on φ (or f), N, T, and molecular architecture. Most molecular theories of polymer-polymer phase behavior have been developed by assuming a simple form for X. Nevertheless, application of these derivations usually requires the use of Equation $X = \alpha T^{-1} + \beta$. Although this complication is ignored for the remainder of this article, the reader should recognize that specific interactions, preferential segment orientation, and equation-of-state effects play an important, albeit poorly understood, role in dictating phase behavior.

Phase state is governed by a balance between enthalpic (H = U + PV where U, P, and V represent the system energy, pressure, and volume, respectively) and entropic (S) factors that together constitute the system (Gibbs) free energy,

$$G = H - TS$$

Theoretical expressions for G are the starting point for predicting equilibrium phase behavior. Statistical models for the molecular configuration of a mixture that depend on molecular architecture, φ (or f), N, and X are the basis for formulating the mixture free energy. Two representative cases, linear binary homopolymer mixtures and diblock copolymers, are examined below.

Linear Homopolymer Mixtures

Equilibrium phase behavior: Nearly 50 years ago Flory and Huggins independently estimated the change in free energy per segment Δ Gm associated with mixing random walk (Gaussian) polymer chains on an incompressible ($\varphi_A + \varphi_B$ =1) lattice,

$$\frac{\Delta G m}{k_B T} = \frac{\phi_A}{N_A} \ln \phi_A + \frac{(1-\phi A)}{N_B} \ln(1-\phi_A) + \phi_A(1-\phi_A)X$$

The first two terms (right-hand side) account for the combinatorial entropy of mixing ΔS_m. Because mixing increases the systems randomness, it naturally increases ΔS_m and thereby decreases the free energy of mixing. Large chains can assume fewer mixed configurations than small chains so that ΔS_m decreases with increasing N. The third term represents the enthalpy of mixing ililm and can either increase or decrease ΔG_m depending on the sign of X. Equation $\frac{\Delta G m}{k_B T} = \frac{\phi_A}{N_A} \ln \phi_A + \frac{(1-\phi A)}{N_B} \ln(1-\phi_A) + \phi_A(1-\phi_A)X$ is a mean-field theory that neglects spatial fluctuations in composition. For N = 1 it reduces to regular solution theory, which is widely applied to low molecular weight solution thermodynamics.

The phase behavior can be predicted with Equation,

$\frac{\Delta G m}{k_B T} = \frac{\phi_A}{N_A} \ln \phi_A + \frac{(1-\phi A)}{N_B} \ln(1-\phi_A) + \phi_A(1-\phi_A)X$ based on the standard criteria for equilibrium, the limits of stability, and criticality evaluated at constant temperature and pressure:

$$\text{Equilibrium}: \frac{\partial \Delta G_m(\phi'_A)}{\partial \phi_A} = \frac{\partial \Delta G_m(\phi'_A)}{\partial \phi_A}$$

$$\text{Stability}: \frac{\partial^2 \Delta G_m}{\partial \phi_A^3} = 0$$

$$\text{Criticality}: \frac{\partial^3 \Delta G_m}{\partial \phi_A^3} = 0$$

where the superscripts' and" refer to separate phases. Plotted in figure is the theoretical phase diagram for the symmetric case $N_A = N_B = N$. The solid curve represents the solution of Equation Equilibrium:

$$\frac{\partial \Delta G_m(\phi'_A)}{\partial \phi_A} = \frac{\partial \Delta G_m(\phi'_A)}{\partial \phi_A}.$$

For combinations of $_xN$ and φ lying inside this curve a mixture separates into two phases with compositions φ'$_A$ and φ"$_A$. Between the solid and dashed curves a homogeneous mixture is thermodynamically metastable, while inside the dashed curve a mixture is thermodynamically unstable. The issues of metastability and stability are discussed in the section below on phase-separation dynamics. At the critical point the equilibrium and stability curves coincides combination of Equations *Stability*: $\dfrac{\partial^2 \Delta G_m}{\partial \phi_A^3} = 0$ and Criticality: $\dfrac{\partial^3 \Delta G_m}{\partial \phi_A^3} = 0$ yields:

$$\phi_c = \frac{N_A^{1/2}}{N_A^{1/2} + N_B^{1/2}}$$

and

$$X_c = \frac{(N_A^{1/2} + N_B^{1/2})^2}{2N_A N_B}$$

where the subscript c signifies criticality. For a symmetric mixture $(XN)_c = 2$ and φ$_c = 0.5$, as indicated in figure below. These calculations illustrate two distinctive features of homopolymer phase behavior. Once a particular monomer pair is chosen, thus fixing $X = f(T)$, the critical temperature T_c can be varied by adjusting N (Eq. $X_c = \dfrac{(N_A^{1/2} + N_B^{1/2})^2}{2N_A N_B}$). Also, the critical composition can be controlled by manipulating the ratio N_A / N_B (Eq. $\phi_c = \dfrac{N_A^{1/2}}{N_A^{1/2} + N_B^{1/2}}$).

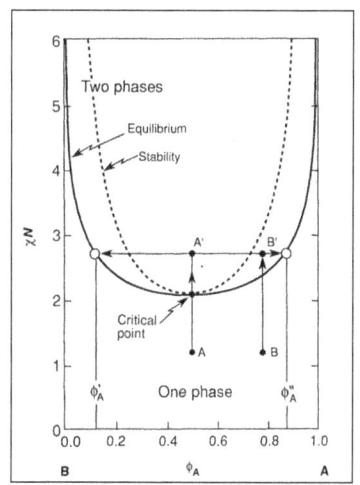

The figure shows the Theoretical phase diagram (mean field) for a symmetric ($N_1 = N_2 = N$) binary mixture of linear homopolymers; N, φ_A and X(T) are the polymer degree of polymerization, volume fraction of component A, and segment-segment interaction parmeter respectively. Inside the equilibrium (solid) curve, two phases exist with compositions φ'_A and φ''_A. In the metastable region (such as point B'), phase separation occurs by a nucleation and growth mechanism, while an unstable mixture (such as point A') spontaneously demixes by spinodal decomposition.

In this way phase diagrams can often be tailored to accommodate experimental constraints, such as glass-transition or thermal-decomposition temperatures. The extent to which this is possible depends on the temperature, composition, and molecular weight dependence of X. For example, if a is positive and ~ is negative in Equation $X = \alpha T^{-1} + \beta$, decreasing temperature always .increases X and an upper critical solution temperature (UCST) results (that is, the two-phase envelope is concave down in the coordinates φ versus T). If a is negative and ~ positive, then a lower critical solution temperature (LCST) may result (that is, the two-phase envelope is concave up in φ versus T) depending on the magnitude of N and β. More complex functional forms for X can produce both UCST and LCST behavior.

In spite of the need to treat X as a largely phenomenalogical parameter, homopolymer mixtures are in many respects more ideal and better applied to fundamental studies of phase behavior than their low molecular weight counterparts. Unlike the latter, this class of mixtures is well described by mean-field theory (such as Equation $X = \alpha T^{-1} + \beta$) over all but a miniscule region of phase space around the critical point. This characteristic can be traced to the large dimensions associated with polymer chains. One useful measure of polymer size is the radius-of-gyration, $R_g = a (N/6)^{1/2}$, where a represents the length of a segment. For polystyrene, a = 6.8 A, which leads to R_g = 270 Å for a sample with a molecular weight of 10^6. In this case a single molecular volume, approximated as $(4/3)\pi R_g^3$, contains 50 other polystyrene chains, which virtually assures average (mean-field) segment-segment interactions except where composition fluctuations become much longer ranged than R_g. Since the mean-field composition fluctuation length ξ scales as (1) $\xi \sim (X_c - X)^{-1/2}$, the condition for crossover from mean-field to non-mean-field behavior ($\xi \sim R_g$), known as the Ginzburg criterion scales as $IX_c - XI - N^{-1}$. Thus, increasing N shrinks the non-me an-field region to a small area around the critical point. In contrast, for monomers (N = 1) composition fluctuations influence the entire phase diagram.

Fluctuation effects are important in many areas of materials science and condensed-matter physics, including magnetism, superconductivity, and liquid crystallinity. However, the ability to influence the Ginzburg criterion through the choice of monomers and N makes polymer mixtures attractive for investigating the limitations of mean-field theory. Recent studies of critical mixtures of poly(vinyl methyl ether) and polystyrene (LCST), and poly(ethylene-propylene) and polyisoprene (UCST) underscore this point in providing the first quantitative verification of the Ginzburg criterion.

Since $X_c - N^{-1}$ (Equation $X_c = \dfrac{(N_A^{1/2} + N_B^{1/2})^2}{2N_A N_B}$), homopolymer phase behavior can be influenced by remarkably subtle chemical or structural variations between components. A dramatic demonstration of this point is provided by mixtures of deuterated and normal (protonated) polymers that are otherwise chemically indistinguishable. Until recently such isotopic mixtures were generally assumed to form ideal solutions (X = 0). However, several years ago it was discovered that above a molecular weight of 1.8×10^5 glmol symmetric mixtures of deuterated and protonated polybutadienes phase separate. This phenomenon has since been shown to be universal and predictable. It derives from the well-known reduction in carbon-hydrogen bond length (-0.1%) that accompanies deuterium substitution in organic molecules. Decreasing the bond length reduces the bond polarizability, which is manifested as a decreased segment polarizability. Together these two effects produce a small positive X parameter, as predicted by Equation $X = \dfrac{3}{16} \dfrac{I}{k_B T} \dfrac{z}{V^2} (\alpha_A - \alpha_B)^2$.

The polymer isotope effect is an extremely valuable tool for investigating the thermodynamics and dynamics of polymer-polymer mixing and demixing. Because polymer isotopes are chemically identical, they very nearly satisfy the assumptions underlying most statistical mechanical theories of polymer thermodynamics, which is that the repeat units are structurally symmetric. This fact is reflected in the ability of Equation $X = \dfrac{3}{16} \dfrac{I}{k_B T} \dfrac{z}{V^2} (\alpha_A - \alpha_B)^2$ to predict the magnitude of X based on independent estimates of a and V. Recent experiments with isotopic polymer mixtures have for the first time identified deficiencies in the Flory-Huggins combinatorial entropy of mixing; prior studies could not discriminate between this effect and various other empirical corrections to X. Other current applications include studies of wetting, interfacial mixing, and phase-separation dynamics.

Significant advances in the theoretical treatment of binary polymer mixtures have also been achieved in the past few years. Freed and co-workers have addressed the issues of structural asymmetry by incorporating a variable monomer structure (such as pendent groups with different sizes and degrees of branching) into an improved lattice theory. This refinement produces a rich concentration dependence to X that varies with the magnitude of €ij (Equation $\epsilon_{ij} = -\sum_{i,j} \dfrac{3}{4} \dfrac{I_i I_j}{I_i + I_j} \dfrac{a_i a_j}{r_{ij}^6}$).

In an alternative approach, Curro and Schweizer report the application of modern liquid-state theory, developed over the past two decades for simple fluids, to the problem of polymer-polymer mixtures. This approach couples conventional long-range polymercoil statistics with the short-range constraints imposed by local liquid structure. They also predict a composition dependent X along with a nonclassical scaling behavior for the critical point; for a symmetric mixture they find X_c - $N^{-1/2}$, in contrast with the Flory-Huggins prediction (E Equation $X_c = \dfrac{(N_A^{1/2} + N_B^{1/2})^2}{2 N_A N_B}$)

that Xc - N^{-1}. Verification of these newly developed theories represents an important experimental challenge.

Phase-separation dynamics: Homopolymer mixtures in the onephase region of the phase diagram are easily homogenized by mechanical mixing or solution casting. Noryl, a commercial plastic composed of poly(phenylene oxide) and polystyrene, is an example of such a homogeneous alloy. Conversely, equilibrium is neither desirable nor practically attainable with two-phase polymer mixtures. Homopolymer phase behavior can be accurately represented by mean-field theory.

Therefore, phase-separation dynamics can be divided into two categories, nucleation and growth and spinodal decomposition, as illustrated in the figure.

Classical nucleation theory predicts that small droplets of a minority phase develop over time in a homogeneous mixture that has been brought into the metastable region of the phase diagram (for example, from point B to B' in figure). Initially droplet growth proceeds by diffusion of material from the supersaturated continuum. However, once the composition of the supernatant reaches equilibrium (φ''_A in figure), further increases in droplet size occur by droplet coalescence or Ostwald ripening; the latter refers to the growth of large droplets through the disappearance ("evaporation") of smaller ones. Because of the extremely low diffusivity ($D - N^{-2}$) and enormous viscosity ($\eta - N^{3.4}$) of polymers, the second stage of growth can be extremely slow and may result in unusual particle-size distributions.

In the metastable state, homogeneous mixtures must overcome a free energy barrier in order to nucleate a new phase. In the thermodynamically unstable state there is no such barrier, and mixtures phase separate spontaneously (for example, from point A to A' in figure). This process, which was first described theoretically by Cahn 25 years ago, is known as spinodal decomposition. It results in a disordered bicontinuous two-phase structure that is contrasted in figure with the morphology associated with the nucleation and growth mechanism. The initial size do of the spinoda structure is controlled by the quench depth Xs - X, where Xs corresponds to the stability limit (dashed curve in figure); deeper quenches produce finer structures. Almost immediately after the bicontinuous pattern begins to form, interfacial tension drives the system to reduce its surface area by increasing d. In symmetric critical mixtures ($N_A = N_B == N$ and $\varphi_c = 1/2$)) coarsening does not disrupt the bicontinuous morphology that evolves through a universal, scale invariant form, as depicted in figure.

The intricate structures associated with spinodal decomposition lead to a variety of interesting materials applications. These include polymer-based membranes, controlled porous glasses, and certain metal and ceramic alloys. Linear homopolymer mixtures have become one of the most attractive systems for studying spinodal decomposition in recent years. Applicability of mean-field theory simplifies the theoretical analysis considerably. However, the greatest advantages are a tunable phase diagram and an extremely low rate of phase separation. Under comparable conditions spinodal decomposition in polymer mixtures is $-N^2$ slower than in low molecular weight mixtures. A recent study of spinodal decomposition in nearly symmetric polybutadiene isotopes illustrates these points. Selection of $N = 3 \times 10^4$ for this pair of isotopes resulted in a UCST at 61 °C. Following a quench from 75° to 25 °C (this is analagous to A to A' in figure), an initial spinodal pattern with d $0.23 \cong \mu m$ developed within 30 min. The structure required 24 hours to coarsen to $1 \mu m$, and 2 months passed before a period of 5 μm was realized. In contrast, in most monomeric mixtures this entire process would take only several seconds. By telescoping the time scale for spinodal decomposition, the evolving composition patterns can be examined in greater detail through the use of scattering (light and neutrons) and microscopy techniques. The results of these studies are enhancing our general understanding of phase-separation dynamics.

Diblock Copolymers

1. Order and disorder: The ensemble of molecular configurations that produces the minimum overall free energy G represents the equilibrium state in a block copolymer melt. Note that

this criterion differs from the condition for equilibrium in homopolymer mixtures (Equation

$$\text{Equilibrium}: \frac{\partial \Delta G_{\mathrm{m}}(\phi'_{A})}{\partial \phi_{A}} = \frac{\partial \Delta G_{\mathrm{m}}(\phi'_{A})}{\partial \phi_{A}})$$ because block copolymers are single component systems

that cannot phase separate. When X > 0 a decrease in A-B segment-segment contacts in an A-B di-block copolymer reduces the system enthalpy H. This process can occur locally, segregating A and B blocks. Segregation is opposed by the associated loss in system entropy that derives from (i) localizing block-block joints at interfaces and (ii) stretching the chains in order to maintain a uniform density (stretching a polymer chain reduces its configurational entropy). Diblock copolymer entropy also scales as S-N^{-1}, and as before it is the product XN that controls the state of segregation. However, the nature of block segregation is quite different from phase separation that occurs with homopolymers.

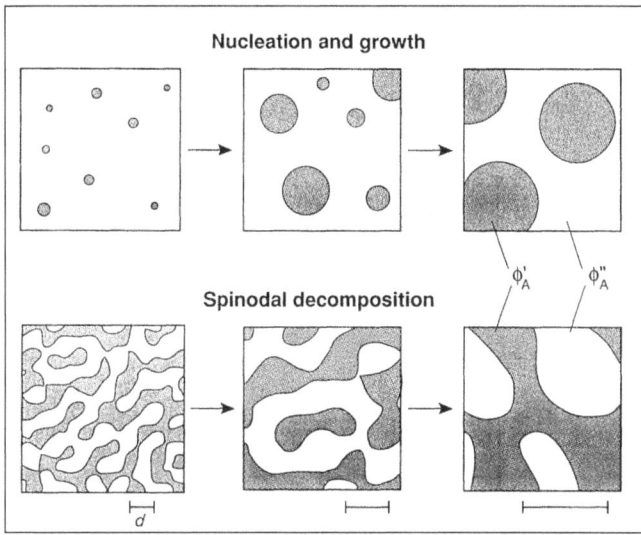

Time evolution of structure in phase-separating binary homopolymer mixtures. Nucleation and growth results when a homogeneous mixture is thrust into the metastable region of the phase diagram (such as B → B' in above fgure). Spinodal decomposition occurs when a mixture is placed in a thermodynamically unstable state (such as A → A' in above fgure). The driving force behind coarsening in both cases is the minimization of interfacial tension through a reduction in interfacial area.

For XN < < 10, entropic factors dominate and diblock copolymers exist in a spatially homogeneous state. Increasing N or X shifts the free energy balance and leads to the development of local composition fluctuations on a scale proportional to the polymer radius-of-gyration ($R_{g}^{2} = R_{g,A}^{2} + R_{g,B}^{2}$), as depicted in the above figure. When XN ≈ 10, a delicate balance exist between entropic and energetic effects. Increasing this parameter further induces a first-order transition to an ordered state. In this case entropically favored but energetically costly curved and disordered microstructures are exchanged for a periodic mesophase. This phase transition, known as the order-disorder transition (oDT), resembles the familiar freezing transition in low molecular weight systems, such as water at O °C. Increasing XN still further leads to sharper microdomain boundaries as the number of A-B segment-segment contacts decrease at the expense of additional chain stretching. In the limit XN > > 10, energetic factors dominate and the ordered microstructures are characterized by narrow interfaces and nearly flat composition profiles.

Thus far we have only considered the symmetric case f = 1/2. Changes in primarily affect the shape and packing symmetry of the ordered microstructure and, except near the ODT, are almost uncorrelated with X^N. Reducing or increasing f places unequal packing and chain stretching constraints on each side of a diblock copolymer molecule and leads to new ordered-phase symmetries over specific ranges in composition. Seven ordered phases have been identified in the polystyrenepolyisoprene (PS-PI) system, as illustrated in the figure below. Where f_s denotes the volume fraction of polystyrene.

For f_s < 0.17, microspheres of polystyrene are ordered on a body-centered cubic lattice in a matrix of polyisoprene. Increasing the composition to 0.17 < f < 0.28 leads to hexagonally packed (hex) cylindrical microdomains. An ordered bicontinuous double diamond (obdd) polystyrene microstructure embedded in a continuous polyisoprene microphase appears where 0.28 < f < 0.34. This most recent addition to the family of documented equilibrium ordered phases in PS-PI is particularly interesting since it contains a triply periodic constant mean curvature (CMC) surface. Because variations in surface area and curvature directly influence the ordered -state free energy, block copolymers can provide a physical manifestation of abstract mathematical concepts used to describe CMCs. Between fs = 0.34 and 0.62 these materials display a lamellar microstructure. Increasing f_s further leads to the corresponding inverted ordered phases. Combining the effects of varying $_xN$ and results in the PS-PI diblock copolymer phase diagram shown in the figure below.

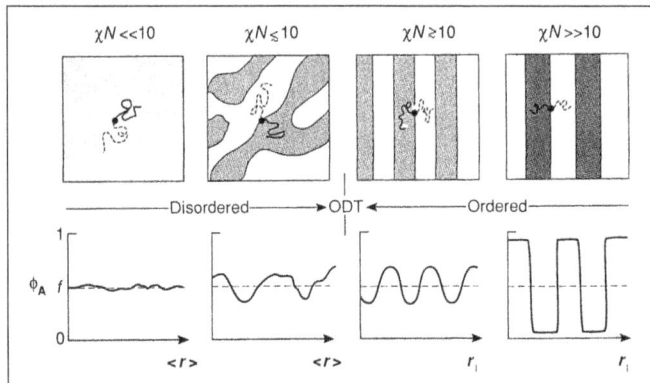

The Evolution of structure with the combined parameter XN for a symmetric, diblock copolymer with f = 0.5. When xN -10, small variations in system entropy (- N^{-1}) or energy (-x) leads to ordered (XN ≥ 10) or disordered (XN ≤ 10) states. A homogeneous composition profile ϕ_A versus r) results when entropic factors dominate (XN < < 10), whereas a strongly microphase segregated pattern characterizes the limit where energetic factors prevail (XN > > 10).

Theories that deal with block-copolymer phase behavior can be divided into two categories: (i) strong segregation limit (SSL, $_xN$ > > 10) and (ii) weak segregation limit (WSL, $_xN$ ≤ 10). In the SSL theories well-developed microdomain structures are assumed to occur with relatively sharp interfaces and chain stretching is explicitly accounted for. The WSL theories are premised on lower amplitude sinusoidal composition profiles and unperturbed Gaussian coils, that is, they neglect chain stretching. Mean-field SSL theories are reasonably successful at predicting how microdomain symmetry, size, and periodicity depend on N, f, and X away from the ODT, although to date the existence of an obdd phase has not been accounted for.

Near the ODT the situation is more complex. Here the delicate balance between X and N can be influenced byf and can lead to curvature in the order-order transition lines. Composition fluctuations, which are insignificant when $_xN << 10$ and $_xN >> 10$, can undermine the use of mean-field theory and further complicate the analysis. Recent theoretical and experimental progress in this area has demonstrated that these effects lead to non-universal phase behavior, that is, a unique phase diagram can be associated with each value of N. However, WSL fluctuation theory fails to account for polymer chain stretching near the oDT, which has been documented experimentally.

Recent experiments on an f = 0.65 poly(ethylene-propylene)-poly(ethylethylene) (PEP-PEE) diblock copolymer have revealed three distinct ordered phases, along with a disordered phase, that can be accessed by varying temperature. This discovery lends support to the notion of a non-universal diblock-copolymer phase diagram; note that more than two decades of intense study of the PS-PI system has yet to uncover any order-order transitions controlled by temperature alone. The increase in phase complexity in PEP-PEE polymers may be attributable to a smaller X parameter (approximately one-fifth that of PS-PI). Decreasing X, which necessitates increasing N in order to maintain an experimentally viable ODT temperature, necessarily increases the sensitivity of a block copolymer system to minor perturbations in free energy, whether energetic or entropic in origin. As a consequence, small variations in interfacial structure or chain stretching might be balanced by minimizing interfacial curvature or spatial variations in micro domain dimensions through transitions between nearly degenerate phases. Hence, block copolymer melts may represent ideal substrates for investigating the concept of periodic surfaces of prescribed mean curvature.

2. Long-range order: At equilibrium an ordered block copolymer will be macroscopically oriented, analogous to a single crystal of low molecular weight material. However, when quenched below the ODT an undisturbed disordered melt will rapidly order [except very close to the ODT] without a preferred direction. This process inevitably leads to short-range order but long-range isotropy, just as quenching water well below the freezing point results in a polycrystalline state. Long-range order can be recovered in such systems by applying a symmetry-breaking field. For example, application of a magnetic field to a ferromagnetic material as it is cooled below the Curie temperature induces a permanent net magnetic moment. Thermal gradients are used to prepare large single crystals of silicon by directional solidification, and an electric field can produce a high degree of orientation in liquid crystalline materials.

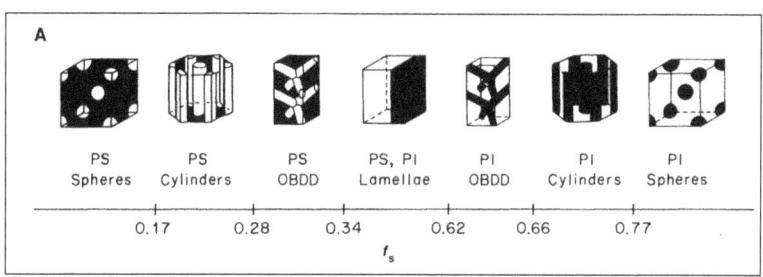

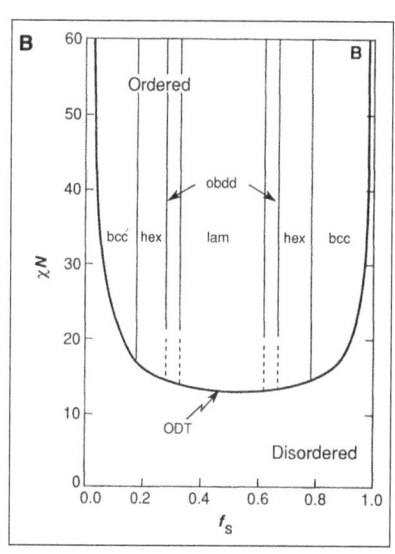

(A) Effect of varying composition on the ordered-phase symmetry in polystyrene-polyisoprene (PS-PI) diblock copolymer; I, refers to the overall volume fraction of polystyrene. (B) Phase diagram for polystyrene-polyisoprene (PS-PI) diblock copolymers. Ordered phases correspond to those illustrated in (A).

Ordered block copolymers respond to mechanical deformation. The application of a large-amplitude oscillatory shear field can convert a quenched (polycrystalline) specimen with f ~ 0.5 into a set of highly oriented lamellae; in this case the lamellae lie parallel to the plane of shear. Cylindrical microstructures also respond to a shear field, resulting in hexagonal single crystals. Such highly oriented materials are useful for studying the equilibriurn state, particularly when multiple ordered phases are possible.

Long-range-ordered block copolymers also exhibit highly anisotropic properties. Because polystyrene is a glass and polyisoprene is rubbery at room temperature, an oriented PS-PI material containing polystyrene cylinders will be rubbery and glassy-like along directions perpendicular and parallel, respectively, to the micro domain long axis. Highly oriented lamellae will also be characterized by anisotropic properties, such as gas permeability. At present, long-range order in block copolymers remains largely a laboratory curiosity. Nevertheless, the ability to control the size, spacing, and materials that constitute highly oriented microstructures such as those illustrated in figure should result in creative applications.

Polymer-polymer phase behavior is a complex interdisciplinary subject that makes contact with numerous basic scientific concepts while also finding innumerable engineering applications. By controlling molecular architecture through synthetic chemistry a single pair of monomers can be directed into a wide variety of multiphase structures that vary dramatically in scale, form, and degree of order. The extremely long molecular relaxation times associated with undiluted high molecular weight polymers can be exploited to probe the fundamental principles of phase separation (such as spinodal decomposition) or ordering dynamics or can be applied toward optimizing physical properties such as toughness.

In this topic an attempt has been made to identify the basic factors that govern polymer-polymer phase behavior by focusing on two representative model molecular architectures. Both linear homopolymer mixtures and diblock copolymers have been subjected to intense study for many years. Nevertheless, the understanding of these most basic of molecular configurations continues to evolve, and this knowledge continues to have a significance impact on science and technology. Furthermore, this subject matter, which draws upon a host of traditional scientific and engineering disciplines, has only begun to be explored. More complex phase behaviors brought about by new molecular architectures, or mixtures of those presently available, should provide enticing challenges and opportunities well into the next century.

Phase Behaviour of Polymer Blends

Contemporary view of Blend Phase Behavior

Blends with a homogeneous amorphous phase exhibit a single glass transition. This transition occurs at a temperature (T_g) intermediate between the values for the two pure component polymers. This intermediate temperature reflects the mixed environment in which the two types of chain segments coexist. This transition is similar to what occurs in a random copolymer. When this homogeneity exists for all blend proportions, the T_g relationship will be something like that shown in figure; and, if this system is an equilibrium state, complete liquid-liquid miseibility exists. If one of the components is crystallizable, a separate crystalline phase of pure 2 can form at temperatures sufficiently below the melting point, T_m, of pure 2. The crystals of 2 will coexist with a mixed amorphous phase made up of components 1 and 2 in the manner pictorially represented in figure. As a result of the depletion of component 2 from the amorphous phase, the shape of the T_g -composition relationship will be affected because the amorphous phase will be richer in species 1 than the overall composition of the blend. Because polymers do not crystallize completely, the amorphous phase will always contain some of the crystallizable species. The fraction of the crystalline phase, based on total blend mass, may be similar to one of the curves shown in the lower part of figure. Formation of miscible blends will affect the temperature interval available for crystallization, T_m - T_g as shown in figure, and, thus, influence the kinetics of this process. Curve A is believed to be the limiting situation for long crystallization times and corresponds to a fixed level of crystallinity that is based on the mass of species 2 in the blend and equal to the crystallinity existing in the unblended polymer. Curves B and C are representative of the crystallinity when finite crystallization conditions are imposed. Total suppression of crystallinity may occur for some composition regions, depending on the time allowed for the process to occur and on the value of the glass transition of polymer 1. Except near T_m, the suppression of crystallization is kinetic in origin because, as shown in figure, the relative magnitudes of the heats of mixing and crystallization in the typical system are such that thermodynamics cannot preclude the formation of crystals of polymer 2 at equilibrium. The melting point of these crystals at equilibrium will be depressed by a few degrees because of the lowered activity or chemical potential of this species in the amorphous phase.

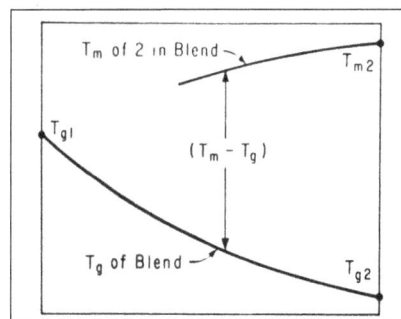 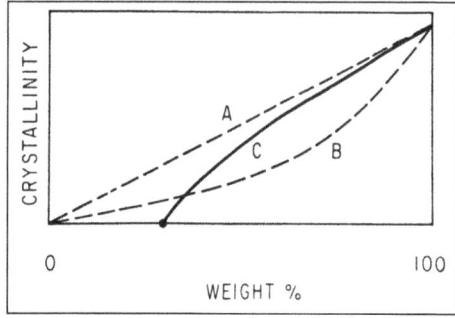

Schematic of transitional and crystallization behavior of miscible blends with one crystallizable component.

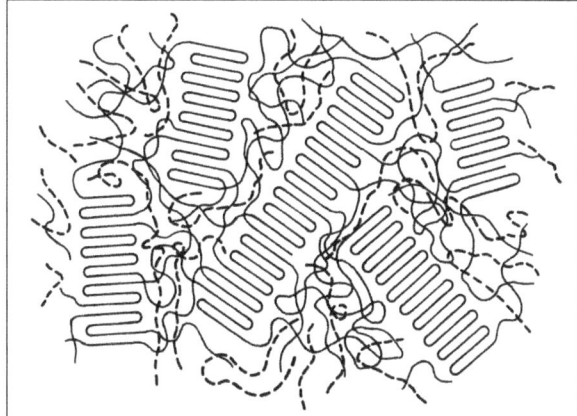

Idealized representation of a blend with a mixed amorphous phase
and with one component partially crystallized.

Mixtures that are homogeneous at one temperature may undergo a liquid-liquid type phase separation at other temperatures. The common experience is for phase separation to occur on cooling, in which case an upper critical solution temperature, UCST, exists. UCST behavior is characteristic of endothermic mixing and a positive entropy of mixing and is well known in mixtures of low molecular weight species and for polymer solutions. However, for blends of high molecular weight polymers, UCST behavior does not seem to occur and generally should not be expected. On the other hand, phase separation on heating caused by a lower critical solution temperature, LCST, is rather prevalent in all blends. LCST behavior is characteristic of exothermic mixing and a negative excess entropy. It is not explained by the simple Flory-Huggins theory; however, the newer "equation of state" theories for mixtures do predict such behavior. The common belief now is that "free-volume" effects are responsible for this mode of phase separation. The LCST is strongly influenced by the strength of the segmental interactions, in addition to free-volume effects. Mixtures with a large negative interaction parameter, B, will not phase separate except at very high temperatures that may exceed the thermal stability of the components.

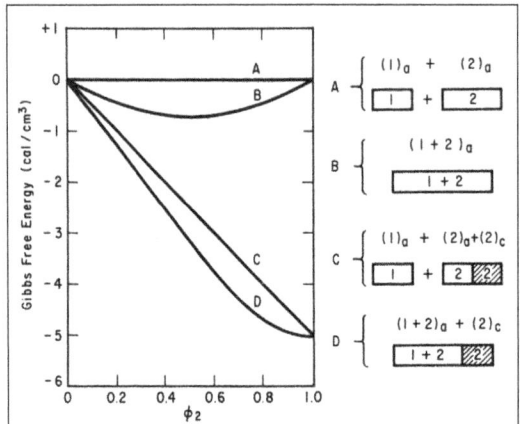

Free energy for miscible blends with and without crystallinity for typical energy parameters.
Notation on right pictorially defines state considered: subscript c and shading denote
crystalline phase and subscript a and lack of shading denote amorphous phase.

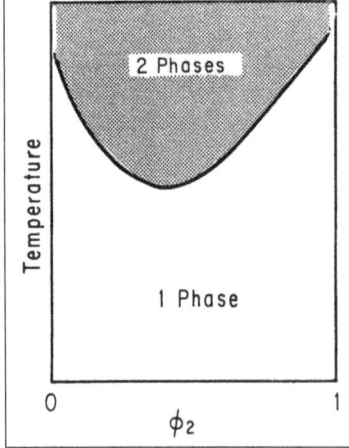

Liquid-liquid phase diagram showing LCST behavior.

Of course, liquid-liquid phase separation temperatures may occur in the vicinity of solid-liquid transitions, T_m or T_g, and make the blend phase diagram quite complex. In such systems, a homogeneous amorphous phase may be observed only over a portion of the composition spectrum.

Intense research efforts are in progress at several laboratories to quantify the interactions for polymer-polymer systems and to understand their origin. Only through such knowledge can equilibrium phase behavior be explained and related to component molecular structures so that technologically important multicomponent polymers can be precisely engineered. Through the use of a variety of techniques, progress is being made to understand the nature of the specific intermolecular and the intramolecular interactions involved. Interestingly, random copolymers often form miscible blends with other polymers, whereas the respective homopolymers do not [e.g., poly(methyl methacrylate) (PMMA)-poly(styrene-co-acrylonitrile), poly (vinyl chloride) (PVC)-poly(ethylene-co-vinyl acetate), or PVC-poly(butadiene-co-acrylonitrile)]. Intramolecular interactions are believed to be a factor in such behavior.

In spite of the intense activity and progress in the area of miscible blends, most polymer-polymer pairs are not miscible to any substantial degree, as the early literature suggested. However, such systems may nevertheless be of great technological value; hence, there is a strong incentive to develop a better scientific understanding of the important issues in such mixtures. Progress has been made in this area, but generally the subject has not yet generated the same fervor as miscibility. The main factors important to immiscible blends are the spatial arrangements of the phases (morphology) and the nature of the interface between them.

The most commonly expected phase morphology for immiscible blends is for one of the components to form a continuous matrix within which the other component is dispersed. The dispersed phase may be nearly spherical particles or highly elongated fibrils depending on the conditions used to fabricate the blend (e.g., elongational flows associated with orientation of sheet or film or with injection molding favor fibrillar morphologies). Generally, the major component is expected to form the continuous phase. However, as suggested in figure, phase continuity is also influenced rather substantially by the relative rheological characteristics, viscosity to a first approximation, of the components. Under certain conditions cocontinuity of both components may exist. Interpenetrating networks, or IPNs, of phases formed from thermoplastic components are to be contrasted from molecular networks formed from two thermosetting polymers described by the same terminology. An IPN morphology is especially useful when the two phases do not adhere well to each other.

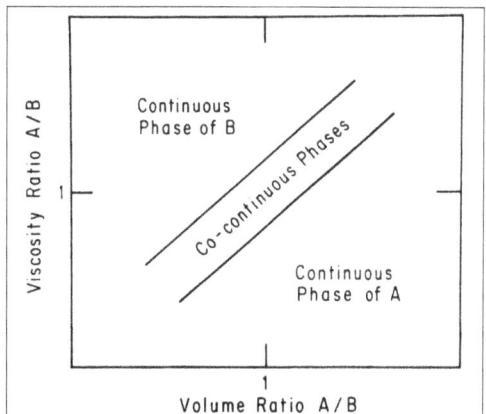

Effect of relative component proportions and viscosity on phase morphology.

Phase morphology is not an inherent characteristic for a blend. It is substantially influenced by the method and conditions used to form the mixture. Technological advances are being made in the control of the nature and scale of phase dispersion in many blend products through the judicious use of sophisticated intensive mixing equipment, such as the various types of twin screw devices. Similarly, proper balancing of the rheological properties of the components in the melt state by selection of appropriate molecular weight distribution grades is another important means for successfully tailoring morphology to meet product needs. In view of these considerations, Figure is clearly a highly oversimplified picture and fundamental research on mixing and flow behavior of multiphase polymer systems is needed to guide technological advances in blend compounding.

Another key factor is the nature of the interface between the phases. In the fluid state, the issue is the magnitude of the interfacial tension because it affects the extent of dispersion during mixing. In the solid state, the related issue is interfacial adhesion, which governs transfer of mechanical stresses between phases and how they share in supporting external loads. For components with low affinity for each other, the melt interfacial tension will be high and a fine dispersion of phases will be difficult to achieve. In the solid state, adhesion will be low and poor mechanical properties result. On the other hand, components with a stronger affinity for each other, but not so great as to cause miscibility, will have a low interfacial tension and high interfacial adhesion, both of which favor good property relationships. Recent work has shown, as might be expected, that systems with significant partial miscibility also exhibit strong interfacial adhesion and good mechanical properties. Selection of systems on the "edge of miscibility" appears to be a means of reaping the benefits of phase separation without incurring the disadvantages. Knowing how to molecularly design for this type of system would provide an important technological advance.

Unfortunately, interactions between polymers that otherwise might be blended advantageously are weak, and the nature of the interface poses problems in both the melt and the solid state. One way to resolve this problem is to add suitably chosen block or graft copolymers that can function as coupling agents between the phases. This concept is widely used in the technology of rubber toughening of plastics. It may be used in a more general way to make grossly immiscible components more compatible to form functionally useful blends. This terminology does not imply that such additives render the system miscible, only that the interfacial problem is solved.

The most obvious choice for an interfacial agent or agent to effect compatibility is one where the segments of the polymers forming phases A and B are identical, respectively, with the segments C and D forming the block or graft copolymer. Recent work has focused on the optimal molecular design of such additives. However, choosing systems where A is identical with C and B is identical with D is often rather restrictive. Frequently, the chemistry required to create such blocks or grafts is not known or is too expensive for practical use. Considerable relaxation of these restrictions occurs when one realizes that similar, or perhaps better, results may be had if A is miscible with C and B is miscible with D. In fact, design of coupling agents may be one of the most important uses of the growing knowledge about miscible blends. In the final analysis, it may be sufficient that A adheres to C and that B adheres to D. In reports of the beneficial use of block copolymers in blends, the simplistic interfacial configuration shown in figure is not possible. Apparently, the benefits stem from the ability of the block copolymer to adhere to both components and its propensity to promote an IPN morphology.

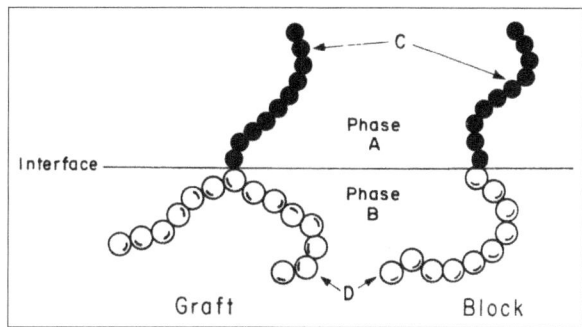

Idealized conformation of a block or graft copolymer at interface between phases.

Property Rehtionships and Applications

The central motive for polymer blending is to create, in an economical fashion, products with desirable properties. Usually, the objective is not to achieve a certain value of a single property but a combination of characteristics (e.g., a critical balance between maximum use temperature, toughness, ease of fabrication, and perhaps resistance to chemicals or burning). A methodical approach to formulation requires experience about how blends behave or, even better, quantitative "mixing rules" for the various individual properties of interest. Parameters of these relationships are blend phase behavior and the interactions between components.

In figure contrasts the behavior of modulus, or stiffness, for miscible and immiscible blends of simple amorphous polymers. Miscible blends behave most nearly like random copolymers in that stiffness decreases precipitously at a single temperature that depends on composition. Thus, miscible blends of poly(phenylene oxide) (PPO) and polystyrene (PS) may be formulated to have any softening temperature between that of the more heat-resistant PPO and that of PS. Of course, the price varies accordingly because PPO is more expensive, but overdesign for a given application can be minimized. Commercial success for this system derives partly from the improved processability resulting from adding PS to PPO. Immiscible blends behave more like composites. In general, properties are dominated by the continuous phase. Commercial blends of acrylonitrile-butadienestyrene (ABS) and polycarbonate (PC), which are not fully miscible, may have, at low stresses, a softening temperature near that of PC when this component is the continuous phase. The cost reduction and improved processability that results from adding ABS to PC has made this system a successful product. When both phases are cocontinuous, stiffness is a nearly additive function of composition. Several products based on this concept made of crystalline polyolefins and ethylene-propylene elastomers have been introduced.

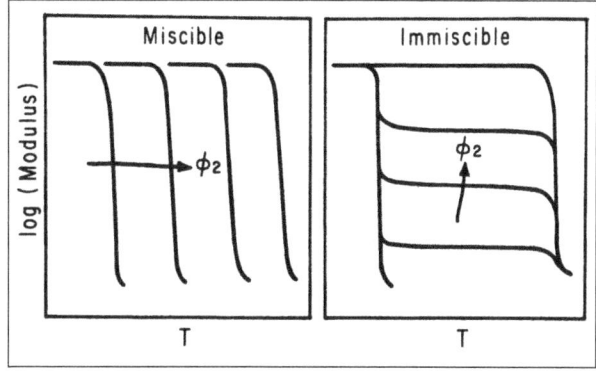

Effect of composition on temperature dependence of the modulus or
stiffness of miscible and immiscible blends.

Failure characteristics, such as strength or toughness, are more complex than stiffness and may show more extreme behavior depending on the blend parameters. Figure illustrates this concept in a schematic way for tensile strength. Curve A shows an interesting synergism that has been observed for both strength and modulus in some miscible blends. This response is believed to be the result of the contraction of free volume, or densification, on mixing that is an expected consequence of the energetic interactions responsible for miscibility in systems like PPO-PS. Curve C illustrates the opposite extreme where grossly immiscible blends fail at low stresses because of poor interfacial adhesion between components. Curve B shows a range of nearly additive responses that might be expected for miscible blends with little densification or for immiscible mixtures with good adhesion.

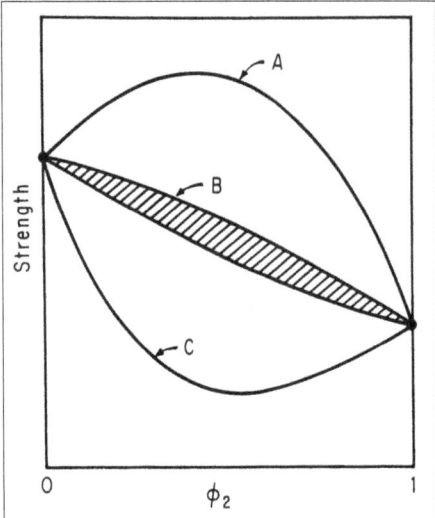

Possible strength-composition relationships for blends.

The figure below shows an example where poor mechanical performance of an immiscible blend has been greatly improved by adding an agent to increase compatibility. A property relationship of type C in figure has been essentially converted to the more desirable type B.

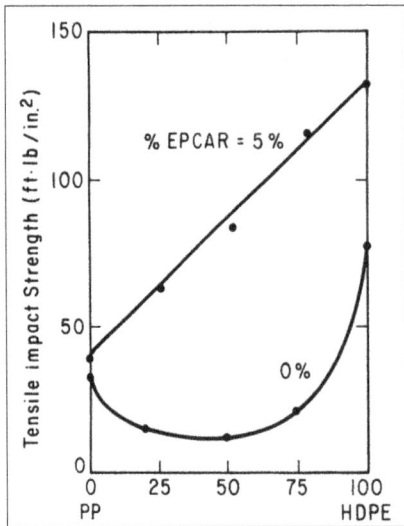

Effect of adding an ethylene-propylene copolymer (Epcar) on the mechanical behavior of polypropylene-high density polyethylene blends.

Interest is growing in the use of polymers as barrier materials and as membranes for separation processes. Blending may provide a new dimension in product development for these applications. Figure shows an example of gas permeation through miscible PPO-PS blends. The experimental data fall well below that predicted by a recently derived mixing rule for this property that assumes additivity of free volume. The discrepancy between the data and the prediction may be a result of the segmental interactions between these two components. Interestingly, addition of small amounts of PS significantly improves the barrier characteristics of PPO. Transport of small molecules in phase-separated blends may be modeled by using composite theory. Phase morphology is an important parameter governing the rate of permeation.

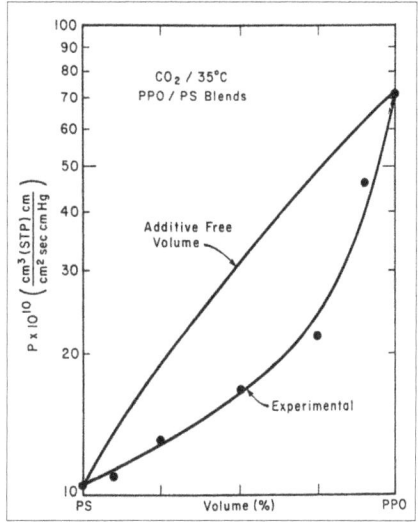

Gas permeation through PS-PPO blends.

Crystallization of one or both components from a miscible blend is an extremely important factor in the resulting property relationship, but space does not permit full elaboration. A good example of the benefits of crystallization is the resistance to chemicals it imparts. For example, blends of semicrystalline poly(vinylidene fluoride) (PVF_2) with amorphous polymethacrylates (PMA) have recently been introduced partly because of the improved chemical resistance PVF_2 brings to the marriage. The same benefits may be imparted by crystallinity in blends with partial or limited miscibility. Recent commercial examples of this type include blends of amorphous PC with either poly(ethylene terephthalate) (PET) or poly(butylene terephthalate) (PBT) (General Electric s Xenoy series).

Basic research on factors affecting the property relationships for polymer blends is expected to become more intense in the near future. The role of the various factors discussed should then become clearer.

Theta Temperature

The Flory temperatures (theta) measured by turbidity experiments performed on gelatin solutions were found to be 12 +/- 0.3, 13 +/- 0.3, 14 +/- 0.3, 14.5 +/- 0.3, and 15 +/- 0.3 degrees C for salt concentrations 0.1, 0.075, 0.05, 0.025, and 0 M (NaCl), respectively. Estimated persistence length (l(p)) of this weakly charged polyelectrolyte could be deduced from the Benoit and Doty relationship with the approximation that this biopolymer assumes a compact near-globular shape at Flory temperature, implying $l(p) = 9(R(h))(2)/(5L(m))$, where $L(m)$ is the contour length and $R(h)$ is the hydrodynamic radius. It was found that $l(p)$ approximately $2.2 +/- 0.2$ nm at room temperature (20 degrees C), invariant of salt concentration. The Flory expansion factor (alpha= $R(h)(T)/R(h)$ (theta) = $1.5+/-0.2$) was found to be almost constant. theta-Composition for this biopolymer was deduced from turbidimitric titration of aqueous gelatin solutions with the alcohols methanol, ethanol, 2-propanol, and tert-butyl alcohol. It appears that hydrophobic interactions play a crucial role in causing chain collapse at theta-temperature and composition.

Polymer Rheology

The word rheology comes from the Greek word "rheos," translated to English as "stream," and it might remind some of the Spanish word "rio." This is important to understand the origin of the word because rheology is the study of the flow (like a stream or a river) and the subsequent deformation of matter as a result of flow. The rheological characteristics of materials directly affect the way that they should be handled and processed. Specifically, the rheological properties determine:

- How the material should be mixed.
- What tools should be used to disperse the material.
- The way that coatings settle.
- The material's shear rate or the rate that the material can be deformed.
- How the material flows into spaces.

These rheological traits are affected by certain inherent qualities of the material, namely:

- Resin, which affects the viscosity and the surface wetting of the material.
- Formulation of the material with its additives.
- Temperature, which directly affects the viscosity of the material.
- Shear, or the way that a material reacts to force.

Understanding all of these traits is fundamental for working with liquids at any level of production. The machinery that is used to disperse and mix liquid materials is related directly to the kind of material and its rheological properties.

Classification of Rheology

Rheology has developed two classes of liquids: Newtonian and Non-Newtonian fluids.

Newtonian Fluids

Newtonian fluids are those which follow Newton's hypothesis, and they are considered to be perfectly viscous. This is because the ratio between the shear rate and shear stress are constant, or in other words, the viscosity of the liquid remains constant at all possible shear rates for a given temperature. Pure water, oils, and organic solvents are all examples of Newtonian liquids. Because of their purity and the lack of dispersion, Newtonian fluids are much easier to measure. Unfortunately, however, they are not very common.

Non-Newtonian Fluids

Most liquids are non-Newtonian fluids, meaning they do not have a constant ratio between their shear rate and shear stress. These fluids can be unpredictable in how their shear stress changes according to the shear rate: as the shear rate increases, the shear stress can either increases or decrease, depending on the fluid's own characteristics. As a result, the viscosity of the material is highly variable. Non-Newtonian fluids will have apparent viscosities that depend entirely on the specific experimental conditions, and when working with these materials, it is important to be completely clear as to what these parameters are.

Non-Newtonian fluids are further classified into two groups: power law fluids and time-dependent fluids.

Power Law Fluids

Power law fluids are categorized based on how the viscosity is affected by the shear. If the viscosity increases as the shear increases, this is a dilatant fluid. Examples of dilatant fluids include candies, sand/water mixtures, and clay slurries.

On the other hand, if the viscosity decreases as the shear increases, this is a pseudoplastic. Pseudoplastics are the most common type of non-Newtonian fluids, and they include inks, mayonnaise, paints, emulsions, and dispersions.

Time-dependent Fluids

The viscosity of time-dependent fluids will change over time. If the viscosity increases as time increases, this is a rheopectic fluid. Fluids that include solvents that evaporate such as adhesives or coatings will fall into this category.

As time increases, however, other fluids will decrease in viscosity if the shear rate is held constant. These are thixotropic, and they can return to their original internal structure before the shear. As such, these materials sometimes give a false high viscosity rate when first measured, because some of the fluid has retained viscosity while in other parts of the fluid, the viscosity has decreased significantly.

Measurement of Rheological Properties

There are four characteristics of fluids that can be measured to determine their rheological properties. These are:

- Viscosity
- Thixotropic Index
- Dispense Rate
- Sag Resistance

Viscosity

Viscosity if often measured via rotating spindle instruments. The instrument will calculated the amount of force (torque) it needs to turn a spindle that is in a sample of the liquid at a specific speed (RPM). The instrument's computer will then use the measured of "internal resistance" of the fluid (measured from the force it needed to apply to turn the spindle) to give the viscosity of the fluid.

Thixotropic Index

The thixotropic index is also known as the Shear Thinning Index (STI), and it is used to determine how stiff a material will be. This is determined by measuring the change in the viscosity (the ratio of shear stress and shear rate) as the shear rate is increased and then decreased. Materials are rated on the index scale from a range of 1 to 5, with one being the most high flow fluids, and five being the least.

Dispense Rate

The dispense rate of materials is measured by changing the pressure, orifice size, and temperature as the material is dispersed. At these conditions are varied, the amount of of the material that is dispersed will vary. This can be measured and compared.

Sag Resistance

Finally, the sag resistance of highly thixotropic products (where the viscosity of the material decreases over time) can be tested as well. This requires a bead of material being applied to a flat surface so that the final flow can be measured.

Interesting behaviour of a material is almost always described by first observing and extracting the qualitative features of the variety of phenomena exhibited. To a novice, this is the place to start. Some questions to ask would be: What makes this phenomena different? How to depict it in terms of a simple model? Is there a "law" that can describe the behaviour? Are there other phenomena that obey similar laws? What role has this played in the state of the universe? Can it be employed for the betterment of quality of life? What are the consequences of this behaviour to processes that manipulate or use the material?

Some of these are simple questions, answers to which may not solve any urgent and present problem. However, it puts it in a larger context. Often, when solving difficult problems in rheology using advanced methods and tools, one tends to miss simple laws exhibited by the materials in reality, knowing which may help in getting an intuitive feel of the solution, and thereby a faster approach to the answer.

What are Polymeric Liquids?

Polymeric liquids are precisely what the name states: namely they are like the liquids we know that flow and have as constituents, some or all being long chain molecules or polymers. The conventional definition used to define simple liquids: that they do not support shear stress at rest, cannot be used to define the liquid state of a polymeric liquid. This is because polymeric liquids, like most other liquids described in this book are complex fluids: They exhibit both liquid and solid like behaviour and some of their dynamic properties may not be thermodynamic constants but some effective constant dependent on the history of forces acting on it. An example is a "viscosity" which is a function of shear rate, or a "viscosity" that changes with time.

Chemical Nature

The common feature of all polymeric liquids considered in this chapter are that they have long chain of monomers joined by chemical bonds. They could even be oligomers or very long chain polymers. Most of them have large molecular weights more than a 1000 and up to about 109. Many materials we use today are polymers or blends of polymers with other materials. In some stage of their processing most of them were in one liquid state or other: as solutions or pure molten form.

Physical Nature

The chief physical property of a polymer that distinguishes itself from other fluids that exhibit complex behaviour is the linearity of the chain. It is not the high molecular weight that leads to the peculiar phenomena but that it is arranged linearly: The length along the chain is much larger than the other dimensions of the molecule. For example, a suspension of polystyrene beads (in solid state) may have high molecular weight per bead. But it may exhibit rheological properties of a suspension rather than a polymer solution. This is because for most properties that we measure, such as the viscosity, it is the beads spherical diameter that matters, and not the molecular weight per bead. There is no linear structure that is exposed to these measurements. On the other hand a solution of polystyrene molecules in cyclohexane is a polymeric liquid.

A huge bowl of noodles is very much like a several molecules of polymers put together. Just as we "see" polymers, so the bowl with these noodles would appear to a Giant. The noodles have the necessary linearity, they are flexible like the polymers are, and as a whole can take the shape of the container they are put into. Can we then use a model bowl full of noodles to understand the behaviour of polymeric liquids. Not entirely. The key difference between the two systems is temperature or random motion. Many of the properties of polymeric liquids we observe, such as that it flows, are due to the random linear translating motion of its constituents, similar as in simple liquids. It is the random motion along the length of the chain and the resulting influence on the other constituents of the material, such as the polymer itself, solvent, or other polymers, that leads to the defining properties of polymeric liquids.

"States" of Polymeric Liquids

The simplest definition (though not a theoretically simpler model to handle) of a polymeric liquid state is the molten state. Take a pure polymer (with no other additives) to a high enough temperature that it is molten. This is like a noodle-state, except that the polymers are in continuous motion. The other extreme of this state is when small amounts of a polymer is added as an additive (solute) to a solvent. This state is called the dilute solution. This is very similar to putting one polymer chain in a sea of solvent. It is equivalent to one polymer because the solution is so dilute that the motion of one polymer does not influence the other polymers in solution. Part of the polymer's motion influences other parts of the same chain. Between these two states of pure and dilute, we can have a range of proportions of the polymer and the other component. Increasing the concentration of the additive polymer from dilute values, we get to the semi-dilute region where the polymers are distributed in the solution such that they just begin to "touch" each other. Further increase in concentration leads to the concentrated-solution where there are significant overlaps and entanglements (like in an entangled noodle soup). The molten state is like a dry noodle with full of entanglements.

Polymer Rheology

Clearly industrial flows are complex, not only because the geometries are complex, but also because the constituents are not usually simple. We have several components in a shampoo, performing various actions. Molecular weight distribution of a polymer is another level of complexity, as polymers are rarely synthesised in a sharp monodisperse population.

This brings in the need to study the behaviour of polymeric liquid in simple flows and for simple systems, with the hope that the knowledge gained can be appropriately used in a complex flow pattern. The word Rheology is defined as the science of deformation and flow was coined by Prof Bingham in 1920s. Rheology involves measurements in controlled flow, mainly the viscometric flow in which the velocity gradients are nearly uniform in space. In these simple flows, there is an applied force where the velocity (or the equivalent shear rate) is measured, or vice versa. They are called viscometric as they are used to define an effective shear viscosity η from the measurment,

$$\eta = \frac{\sigma_{xy}}{\gamma}$$

where σ_{xy} is the shear stress (measured or applied) and $\gamma^{\cdot}$ is the shear rate (applied or measured). Viscosity is measured in Pa-s (Pascal second).

Rheology is not just about viscosity, but also about another important property, namely the elasticity. Complex fluids also exhibit elastic behaviour. Akin to the viscosity defined above being similar to the definition of a Newtonian viscosity, the elasticity of a complex material can be defined similar to its idealised counterpart, the Hookean solid. The modulus of elasticity is defined as:

$$G = \frac{\sigma_{xy}}{\gamma}$$

where γ is called the strain or the angle of the shearing deformation. G is measured in Pa (Pascal). G is one of the elastic modulii, known as the storage modulus, as it is related to the amount of recoverable energy stored by the deformation. G for most polymeric fluids is in the range 10–10^4 Pa, which is much smaller than that of solids (> 10^{10} Pa). This is why complex fluids, of which polymeric fluids form a major part, are also known as soft matter, i.e. materials that exhibit weak elastic properties.

Rheological measurements on polymers can reveal the variety of behaviour exhibited even in simple flows. Even when a theoretical model of the reason for the behaviour is not known, rheological measurements provide useful insights to practising engineers on how to control the flow (η) and feel (G) polymeric liquids. In the following section we sample a few defining behaviours of various properties in commonly encountered flows. In the following sections we present a broad overview of the variety of phenomena observed in polymeric liquids and how the rheological characterisation of the liquids can be made.

Visual and Measurable Phenomena

Some of the striking visual phenomena are associated with flow behaviour of polymeric liquids. These are best seen in recorded videos. We describe a few of them here and provide links to resources where they may be viewed.

Weissenberg Rod Climbing Effect

When a liquid is stirred using a cylindrical rod, the liquid that wets the rod begins to "climb" up the rod and the interface with the surrounding air assumes a steady shape dangling to the rod, so long as there is a continuous rotation. In contrast in a Newtonian liquid, there is a dip in the surface of the liquid near the rod. Rod climbing is exhibited by liquids that show a normal stress difference. In Newtonian liquids the normal stresses (pressure) are isotropic even in flow, whereas polymeric liquids, upon application of shear flow, begin to develop normal stress differences between the flow (τ_{xx}) and flow-gradient directions (τ_{yy}).

Extrudate or Die Swell

This phenomenon is observed when polymeric melts are extruded through a die. The diameter of liquid as it exits a circular die can be three times larger than the diameter of the die, whereas in the case of Newtonian fluids it is just about 10% higher in the low Reynolds number limit. One of the important reasons for this phenomena is again the normal stress difference induced by the shear flow in the die. As the fluid exits the die to form a free surface with the surrounding air, the accumulated stress difference tends to push the fluid in the gradient direction.

Contraction Flow

Sudden contraction in the confining geometry leads to very different streamline patterns in polymeric liquids. In Newtonian liquids at low Reynolds number, no secondary flows are observed. Whereas in polymeric liquids, including in dilute polymer solutions different patterns of secondary flow are observed. These include large vortices and other instabilities. These flows are undesirable in many situations in polymer processing as it leads to stagnation and improper mixing of the fluid in the vortices.

Tubless Siphon

In a typical syphoning experiment, a tube filled with liquid drains a container containing the liquid at a lower pressure, even though the tube goes higher than the liquid surface. When the tube is lifted off the surface of the liquid, the flow immediately stops. But this is the case in Newtonian liquids. In the case of polymeric liquids, the liquid continues to flow with a free surface with the air without the tube, as the tube is taken of the surface.

Elastic Recoil

A "sheet" of polymeric liquids pouring down from a vessel can be literally cut with a pair of scissors. Very similar to a sheet of elastic solid, the top portion of the cut liquid recoils back into the jar.

Turbulent Drag Reduction

In most of phenomena discussed so far the concentration of the polymer was about 0.1% or higher, and the viscosity of such systems are usually large that it is not common to encounter large Reynolds number flows that lead to turbulence. In smaller concentrations of about 0.01% where the solution viscosity is not significantly enhanced above the solvent's viscosity, turbulence can be easily observed. The interesting feature of such turbulent flows, at least in pipe geometries is that the turbulent friction on the walls is significantly less, upto nearly five times. This phenomenon has been used in transportation of liquids and in firefighting equipment.

Relaxation Time and Dimensionless Numbers

One of the simplest and most important characters of polymeric liquids is the existence of an observable microscopic time scale. For regular liquids the timescales of molecular motion are in the order of 10^{-15} seconds, associated with molecular translation. In polymeric liquids, apart from this small time scale, there is an important timescale associated with large scale motions of the whole polymer itself, in the liquid they are suspended in (solutions or melts). This could be from microseconds to minutes. Since many visually observable and processing time scales are of similar order, the ratio of these time scales becomes important. The large scale microscopic motions are usually associated with the elastic character of the polymeric liquids. In the chapter on polymer physics there is a discussion of the relaxation times. The relaxation time is the time associated with large scale motion (or changes) in the structure of the polymer, we denote this time scale by λ.

The microscopic timescale should be compared with the macroscopic flow time scales. The macroscopic time scales arises from two origins. One is simply the kinematic local rate of stretching of the fluid packet (strain rate). This is measured by the local shear rate $\dot{\gamma}$ for shearing flows or the local elongation rate $\dot{\varepsilon}$ for extensional flows. The other is a dynamic timescale associated with the motion of the fluid packets themselves. Examples are the time it takes for a fluid packet to transverse a geometry or a section, pulsatile flow, etc. We denote this timescale by t_d. Except for viscometric flows, the macroscopic timescales may not be known a priori, and have to be determined as part of the solution. For example the nature of the fluid's viscosity could alter the local shear rate $\dot{\gamma}$, or the time it spends in a particular section t_d. To know the this dependence, we need to solve the fluid dynamic equations in the given geometry.

Weissenberg Number

The ratio of the microscopic time scale to the local strain rate is called as the Weissenberg Number:

$$Wi = \lambda\dot{\tilde{a}} \quad \text{or} \quad \lambda\dot{\varepsilon}$$

Note that the strain rate is the inverse of the kinematic timescale. Flows in which the Wi are small, Wi << 1, are in which elastic effects are negligible. Most of the flow effects are seen around Wi ~ O. For large Wi >>1, the liquid behaves almost like an elastic solid.

Weissenberg number is used only in situations where there is a homogeneous stretching of the fluid packet in the flow. That is the strain rates are uniform in space and time. Such a flow is encountered only in viscometric flows and theoretical analysis, as it is hardly observed in any practical application.

Deborah Number

In most practical applications the fluid packets undergo a non-uniform stretch history. This means that they could have been subjected to various strain rates at various times in their motion. Therefore no unique strain rate can be associated with the flow. In these cases it is customary to refer to the Deborah number defined as the ratio of the polymeric time scale to the dynamic or flow timescale:

$$De = \frac{\lambda}{t\,d}$$

For small De <<1, the polymer relaxes much faster than the fluid packet traverses a characteristic distance, and so the fluid packet is said to have "no memory of its state" a few td back. On the other hand for De ~ O, the polymer has not sufficiently relaxed and the state a few td back can influence the motion of the packet now (because this can affect the local viscosity and hence the dynamics).

Table: Scaling of the relaxation time λ of the polymeric liquid with molecular weight M for different classes of polymeric liquids.

Class	Scaling
Dilute solution in poor solvent	$\lambda \sim M^{1.0}$
Dilute solution in θ-conditions	$\lambda \sim M^{1.5}$
Dilute solution in good solvent	$\lambda \sim M^{1.8}$
Semi dilute solution	$\lambda chain \sim M^{2}$
Entangled Melts	$\lambda rep \sim M^{3.4}$

Relaxation Time Dependence on Molecular Weight

Since the time large scale motion of the molecule (i.e. its relaxation time) depends on the linear size of the molecule, the molecular weight of a polymer has a direct bearing on the relaxation times. The dependence on molecular weight is not absolute. That is we cannot say that for a given molecular weight two different polymers will have the same relaxation time. It is only a scaling dependence for a class of polymeric liquids (We can only say that the relaxation time scales with molecular weight power some exponent). This dependence is summarised in the above table. The scaling given for semi-dilute and entangled melts is only indicative of the longest relaxation time; there are several relaxation process in these systems, and the way experimental data is interpreted from measurements carried out at various temperatures.

Linear Viscoelastic Properties

In general the elastic nature of a material is associated with some characteristic equilibrium microstructure in the material. When this microstructure is disturbed (deformed), thermodynamic forces tend to restore the equilibrium. The energy associated with this restoration process is the elastic energy. Polymeric liquids have a microstructures that are like springs representing the linear chain. Restoration of these springs to their equilibrium state is through the elastic energy that is "stored" during the deformation process. But polymeric fluids are not ideal elastic materials, and they also have a dissipative reaction to deformation, which is the viscous dissipation. For small deformations, the response of the system is linear, meaning that the response is additive: effect of sum of two small deformations is equal to the sum of the two individual responses. Linear visco-elasticity was introduced in the chapter on Non-Newtonian Fluids.

Linear viscoelastic properties are associated with near equilibrium measurements of the system that is the configuration of polymers are not removed far away from their equilibrium structures. Most of the models described in the "Polymer Physics" chapter deal with such a situation. Study of linear viscoelastic properties can reveal information about the microscopic structure of polymeric liquids. The term rheology is used by physicists to usually refer to the linear response and by engineers to refer to the large deformation (shear or elongational) or non-linear state.

An important point to note here is that it is just not polymeric liquids that show elastic behaviour, but all liquids do, at sufficiently small time scales. Typical values of relaxation times and the elastic modulus for various liquids is given in table.

Table: Linear Viscoelastic properties of common liquids, values are typical order of magnitude approximations.

Liquid	Viscosity η (Pa.s)	Relaxation time λ (s)	Modulus G (Pa)
Water	10^{-3}	10^{-12}	10^9
An Oil	0.1	10^{-9}	10^8
A polymer solution	1	0.1	10
A polymer melt	10^5	10	10^4
A glass	$>10^{15}$	10^5	$>10^{10}$

Commonly used tests to study the linear response are - Oscillatory Controlled stress/strain is applied in a small amplitude oscillatory motion and the response of the strain/stress is measured. Stress Relaxation A constant strain is applied and the decay of the stresses to the equilibrium value is studied.

Creep A constant stress is applied and the deformation response is measured though all of the above tests can also be carried out in the non-linear regime, as the limits of linear regime are not known a priori, sequence of tests are carried out to ensure linear response.

Zero-Shear Rate Viscosity

The zero shear rate viscosity η_0 is the viscosity of the liquid obtained in the limit of shear rate tending to zero. Though the name suggests that it is a shear viscosity, it is still in the linear response regime because the shear rate is approaching zero. In practise it is not possible to attain very low shear rates for many liquids owing to measurement difficulties. In these cases the viscosity obtained by extrapolating the viscosities obtained at accessible shear rates. The zero shear rate viscosity is an important property to characterise the microstructure of polymeric liquid.

For dilute solutions, since the polymer contribution to the total viscosity is usually small, it is useful to define an intrinsic viscosity at zero shear rate as:

$$[\eta]_0 \equiv \lim_{\dot{\gamma} \to 0}[\eta] \equiv \lim_{\dot{\gamma} \to 0} \lim_{c \to 0} \frac{\eta - \eta_s}{c\eta_s}$$

The intrinsic viscosity scales as:

$$[\eta]_0 \sim \frac{\lambda}{M}$$

where the scaling of λ is given in table for various types of dilute solutions.

In semi-dilute regime, one is more interested in the scaling of the viscosity with concentration. The usual way to report viscosity is through the specific viscosity (and not the intrinsic viscosity):

$$\eta_{sp0} = \eta_0 = \eta_s$$

The specific viscosity scales linearly with concentration in the dilute regime. In the semi-dilute regime, under θ-conditions $\eta_{sp0} \sim c^2$ and in the concentrated (and melt) regime $\eta_{sp0} \sim c^{14/3}$.

The scaling with respect to concentration in good solvents is weaker to about $\sim c^{1.3}$ in semi-dilute and $\sim c^{3.7}$ in the concentrated. More details of this scaling can be found in Ref. The scaling behaviour is summarised in the figure below.

Oscillatory Response

The typical response of a polymeric melt to an oscillatory experiment is shown in figure. The symbols used here are the complex modulus G^* : G' for the storage component (real) and G'' for the loss component (imaginary), as defined in the Non-Newtonian Fluids chapter. G' represents the characteristic elastic modulus of the system and G'' measures the viscous response. At high frequencies the response is glassy (which is typically seen at temperatures around glass transition). The dominant elastic response is seen in the rubbery region where the storage modulus shows a plateau. The plateau region is clear and pronounced in higher molecular weight polymers (with entanglements) in the concentrated solution or the melt states, as shown in figure. In this region the storage modulus (elastic) is always greater than the loss (viscous) modulus. The value of G' at the plateau, is known as the plateau modulus G_N^0, and is an important property in understanding the dynamics of the polymers in the melt state.

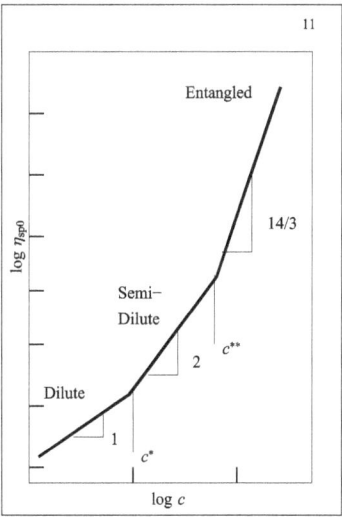

Scaling of the specific viscosity ηspowith concentration of polymers. The first transition denotes the semidilute regime and the second corresponds to the entangled regime.

The low frequency response (or long time response) is always viscous. The viscous regime behaviour is characteristic of all materials (including solids) showing Maxwell behavior where, G is the elastic modulus (constant in the Maxwell model), and η_o is the zeros hear rate viscosity:

$$G' \approx G\lambda^2\omega^2$$

$$G'' \approx \eta_0\omega$$

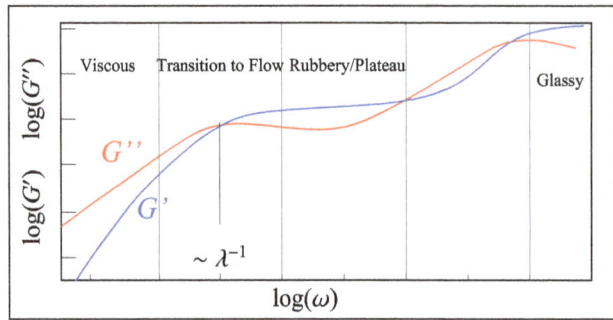

Typical regimes in the complex modulus obtained using an oscillatory response of a polymeric liquid.

This behaviour is also shown by polymeric liquids in the dilute and semi-dilute regimes. The characteristic relaxation time of the structured liquid can be obtained from the inverse of the frequency where G' and G" cross over in the flow transition regime:

$$\lambda = \frac{G'}{G''\omega},$$

where G' and G" are measured in the viscous regime with the linear and quadratic scaling respectively. As seen before, increase in molecular weight increases the relaxation time, so does the increase in concentration. Therefore with increasing concentration or molecular weight of the polymer, the intersection of the two curves shifts more and more towards the left (low frequencies).

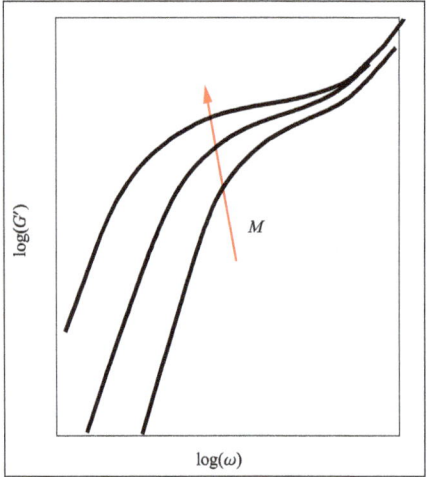

Increase in the plateau region with increase in molecular weight.

The low frequency response of the loss modulus can also be used to obtain the zero-shear rate viscosity ηo from the Equation $G'' \approx \eta_0\omega$.

The low frequency response, close to the plateau region can also be used to distinguish the type of polymer melt. A schematic diagram of the plateau region behaviour of storage modulus G' for various types of polymers is shown in figure. Unentangled melts do not have any plateau region, and directly make the transition to the Maxwell region. Entangled melts show a plateau region. Cross-linked polymers have a wide and predominant plateau region. The transition to the Maxwell behaviour is almost not seen due to limitations in observing very cycles, just as in solids. The link between the plateau region and the cross-linking suggests that the entanglement acts like a kind of constraint (like the cross-links) to the motion of the polymer contour, leading to the plateau region.

Stress Relaxation

The relaxation of the stress modulus G(t) in response to a step strain (for small strains) in the linear regime, is equivalent to the oscillatory response G* (ω), one being the Fourier transform of the other. A schematic diagram of the stress relaxation is shown in figures in linear and logarithmic scale respectively. The initial small time response of G(t) is equivalent to the high frequency response of G*(ω), and the long time response is equivalent to the low frequency response. In small times a polymeric substance shows a glassy behaviour, which goes to the plateau region (seen clearly in the logarithmic scale in figure) and finally to the terminal viscous decay. Since G is related to the elasticity, the response can be understood as being highly elastic at small times, which decays and begins to "flow" at large times: Given sufficiently long time any material flows.

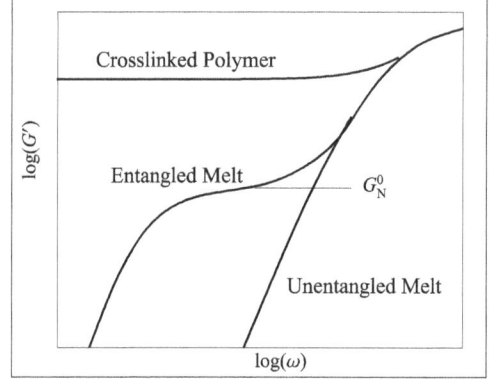
Low frequency response of various type of polymers.

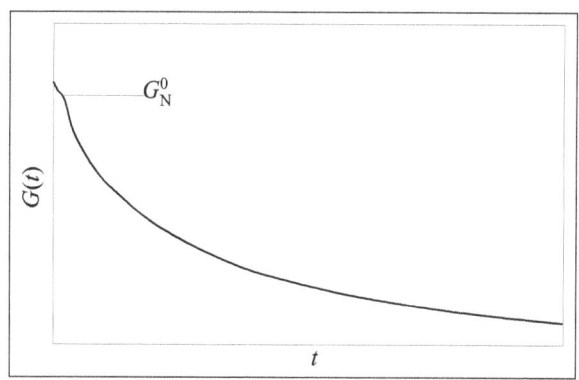
Stress relaxation in response to step strain in linear scale.

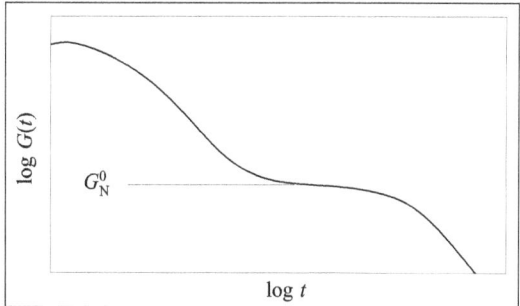
Stress relaxation in response to step strain in logarithmic scale.

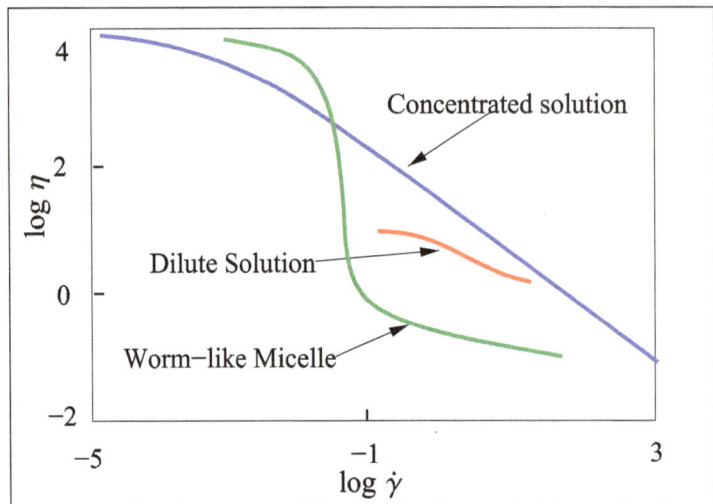

Shear thinning of polymeric liquids is more pronounced in concentrated
solutions and melts than in dilute solutions.

Flow Viscosity and Normal Stress

So far we discussed the behaviour of polymeric liquids slightly disturbed from equilibrium, in the linear regime. Here we show the behaviour in flowing systems that produce significant deviations from the equilibrium microstructure.

Shear Thinning

Most polymeric liquids have their effective viscosity reduced upon shearing. The viscosity defined

in Equation $\eta = \dfrac{\sigma_{xy}}{\gamma}$ is a decreasing function of the shear rate. Figure shows the decrease in the

viscosity for two polymeric liquids. The shear thinning is not so much pronounced in dilute solutions as it is in concentrated solutions and melts.

A special class of polymers is known as Living Polymers, which are long linear structures formed from cylindrical liquid crystalline phases of micelles. They are called living polymers because they form and break along their length owing to thermodynamic and flow considerations. They are also known as worm like micelles. At equilibrium, their behaviour is very similar to a high molecular weight concentrated solution of polymers: large viscosity and elastic modulus. However upon shearing they can break leading to the behaviour exhibited by low molecular weight counterparts, aligning in the shearing direction. The viscosity reduction upon shearing is therefore very significant and sharp in these systems, as depicted schematically in figure. Examples of everyday use are shampoo and shower gels which are very viscous at rest, but can be easily flown out of the container by gravitational forces (which are sufficient to overcome the viscosity after they begin to flow).

Normal Stresses

The normal stress difference is zero for a liquid that is isotropic. Polymeric liquids having microstructure can develop anisotropy in the orientation of the constituent polymers in flow, there by leading to normal stress differences. The normal stress behaviour shows a similar behaviour to that of shear stress. The normal stress difference has two components:

$$N_1 = \tau_{xx} - \tau_{yy}$$

$$N_2 = \tau_{yy} - \tau_{zz}$$

where for planar Couette flow, x is the direction of flow, y is the direction of the gradient and z is the vorticity direction. Similar to the viscosity, which is a coefficient of the shear stress, we can define two coefficients for the normal stresses: the only difference is the denominator which is $\dot{\gamma}^2$, because it is the lowest power of the shear rate that the normal stresses depend on. The coefficients are called as First normal stress coefficients, Ψ_1 and Ψ_2 defined as:

$$\psi_1 = \frac{N_1}{\dot{\gamma}^2}$$

$$\psi_2 = \frac{N_2}{\dot{\gamma}^2}$$

The normal stress coefficients show shear thinning very similar to the viscosity, however for very large shear rates, the absolute value for the normal stress difference can become larger than the shear stress as shown in figure. Such behaviour is seen in concentrated solutions and in melts. The second normal stress N_2 is usually zero for polymeric liquids.

Elongational Flow Viscosity

Extensional or elongational flow where the local kinematics dictates that the fluid element is stretched in one or more directions and compressed in others. Liquids being practically incompressible, the elongational stretch conserves the volume. Elongational flows are encountered many situations. Though there are very few systems where it is purely extensional, there are several cases where there is significant elongation along with rotation of fluid elements, the two basic forms of fluid kinematics due to flow. Sudden changes in flow geometry such as contraction or expansion, spinning of fibres, stagnation point flows, breakup of jets or drops, blow moulding, etc.

The simplest elongation is called as the uniaxial elongation the velocity gradient tensor for which is given by:

$$\nabla v = \dot{\varepsilon} \begin{bmatrix} 1 & 0 & 0 \\ 0 & -\frac{1}{2} & 0 \\ 0 & 0 & -\frac{1}{2} \end{bmatrix}$$

which corresponds to stretching along the x direction and equal compression along y and z directions. $\dot{\varepsilon}$ is called the elongation or extension rate. The elongational (or tensile) viscosity, is defined as:

$$\eta_E = \frac{\sigma_{xx} - \sigma_{yy}}{\dot{\varepsilon}} = \frac{\sigma_{xx} - \sigma_{zz}}{\dot{\varepsilon}}$$

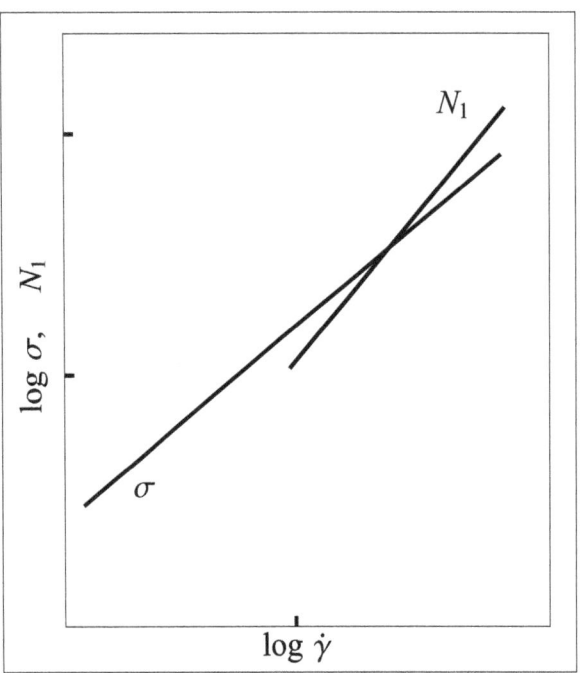

Comparison of the absolute values of normal stresses with that of the shear
stresses as a function of shear rate for concentrated solutions and melts.

The elongational viscosity, like the shear viscosity is a function of the shear rate. However, in the case of elongational flows, it is difficult to measure the steady state value. In experiments it is only possible to access a time dependent value $\eta_E^+(t, \dot{\varepsilon})$ which is a transient elongational viscosity or more precisely the tensile stress growth coefficient. The elongational viscosity is defined as the asymptotic value of this coefficient for large times t → ∞. The behaviour of the transient growth co-efficients for various elongation rates is shown in figure below. The elongational viscosity abruptly increases to a high value in short times. This phenomenon is called as strain hardening. The elongational strain is measured by the Henky strain defined as:

$$\varepsilon = \dot{\varepsilon}t = \log\left(\frac{L(t)}{L_0}\right)$$

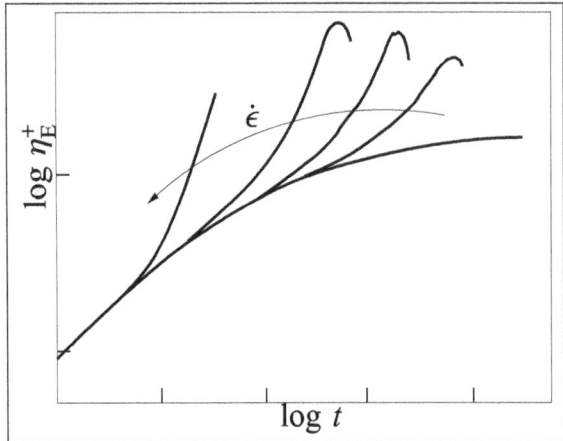

Transient growth coefficient (trbansient elongational viscosity) for increasing strain rates $\varepsilon^{\cdot}$.

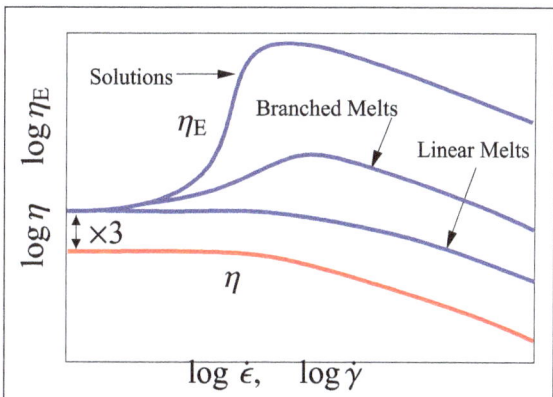

Behaviour of the elongational viscosity with elongation rate for various polymeric liquids.
The elongational viscosity is not necessarily the steady state value, and it could be
the maximum or the value at the terminal Hencky strain.

where L_0 is the initial length of an element along the stretch direction and L(t) is the deformed length that grows as $L(t) = L_0\, e^{\dot{\varepsilon}t}$ The elongational viscosity attains a maximum and then falls down.

It is not possible to easily measure the terminal or asymptotic value of the viscosity. This is because for a Hencky strain of $\varepsilon = 7$, the elongation required is about 1100 times the initial value. It is a convention therefore to report the maximum value of the transient tensile growth coefficient η_E^+ as a function of the strain rate for practical applications. The "Steady values" of the viscosity reported are either the maximum values at a given strain rate, or the value of the coefficient at a given experimentally accessible Hencky strain. The behaviour of this viscosity is shown in figure for various polymeric liquids.

Another useful way to report the elongational viscosity is the Trouton ratio, which is defined as:

$$T_R = \frac{\eta_E(\varepsilon)}{\eta(\sqrt{3}\dot{\varepsilon})}$$

where the shear viscosity is measured at a shear rate $\dot{\gamma} = \sqrt{3}\dot{\varepsilon}$. For Newtonian (or inelastic liquids) the elongational viscosity is three times the shear viscosity. For polymeric liquids, at low shear (and elongation) rates, the Trouton ratio T_R is always $T_R \approx 3$. The Trouton ratio plots, such as the one shown in figure provide an indication of the extent of elongational viscosity effects in relation to the shear viscosity. The most dramatic effects are in dilute solutions of polymers where, depending on the molecular weight of the polymer the Trouton ratio can be several orders of magnitude higher than unity.

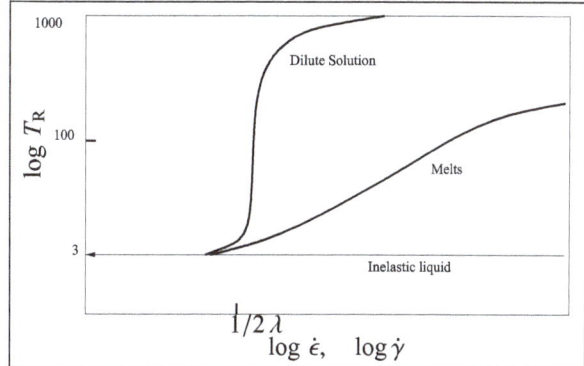

Trouton ratios for various polymeric liquids. Dilute solutions show
the most dramatic increase in the elongational viscosity.

Modelling and Physical Interpretation

Modelling in polymer rheology is mainly of two types: Phenomenological and molecular modelling. Some general phenomenological models have been discussed in the chapter on Non-Newtonian Fluid Mechanics. The basic molecular models have been discussed in the chapter on Polymer Physics: the Rouse and Zimm models. Here we present some simple physical interpretations of the polymer rheological behaviour.

Polymer Solution as a Suspension

The simplest explanation of the polymeric liquid viscosity is that of a dilute solution. A dilute solution is like a suspension of colloidal particles, except that the particles are not spherical. The polymers in equilibrium have a coil like structure. Considering the polymer coils to represent colloidal particles, the viscosity can be related to the volume fraction. For dilute colloidal suspension the viscosity of the suspension is more than the solvent viscosity, and is given by the Einstein expression:

$$\eta = \eta_s\left(1 + \frac{5}{2}\phi\right)$$

where φ is the volume fraction of the colloidal particle. This approximation is valid in the limit $\varphi \to 0$, for spherical particles. Polymer coils are not rigid spheres, but behave more like a porous sphere (on the average, because the shape of the coil changes continuously due to thermal Brownian forces from solvent). Because of this the increase in viscosity is reduced. For dilute polymers the viscosity can be written as:

$$\eta = \eta_s(1 + U_{\eta R}\phi)$$

where $U_{\eta R}$ is a constant $< 5/2$. This constant is an universal constant for all polymers and solvents chemistry and depends only on the solvent quality (which could be poor, theta, or good solvents). The Zimm theory predicts this constant to be 1.66, whereas experiments and simulations incorporating fluctuating hydrodynamic interactions (without the approximations made in Zimm theory) observe the value to be $U_{\eta R} = 1.5$.

Shear Thinning

The extent of shear thinning is much more significant in concentrated solutions and melts in comparison with dilute solutions. The explanation for viscosity decrease in dilute solutions is subtle and involves several factors. We will restrict to simple explanations in entangled systems (i.e. in concentrated solutions and melts). In the entangled state, the microstructure of the solution is like a network: the entanglements act like nodes of a covalently bonded network junctions which restricts the motion of the polymers and therefore of the solution. This is the reason for the high viscosity of the solution. However, upon shearing the chains begin to de-entangle and align along the flow. This reduces the viscosity.

Normal Stresses

The existence of normal stresses in polymeric systems is due to the anisotropy induced in the microstructure because of flow. This can be easily understood in the case of dilute polymer solutions. In the absence of flow, the coil like structure of the molecule assume a spherical pervaded volume, on the average. The flow causes the molecules to stretch towards the direction of flow and tumble. This results in an pervaded volume that is ellipsoidal and oriented towards the direction of flow. The restoring force is different in the two planes xx and yy. This results in the anisotropic normal forces.

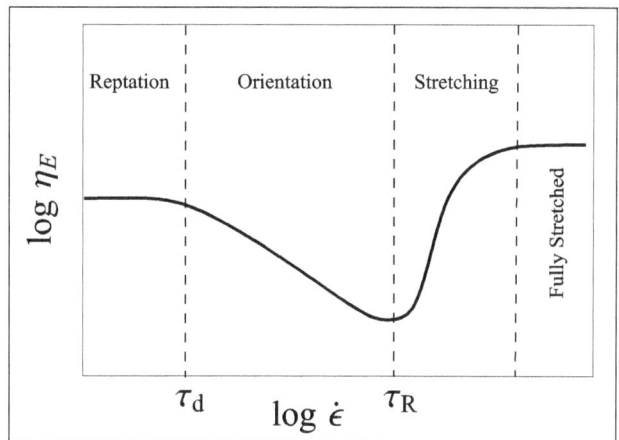

Theoretical prediction of the elongational viscosity in the "standard molecular theory" or the tube model, for linear polymer melts.

Elongational Viscosity

As shown in figure dilute solutions have a dramatic effect on the elongational viscosity. The main reason for this is in the dilute state at equilibrium, the polymers are in a coil like state in a suspension whose viscosity is only slightly more than the solvent viscosity. Whereas upon elongation the chains are oriented in the direction of elongation. After a critical elongation rate given by the Weissenberg number Wi = 0.5, the thermal fluctuations can no longer hold against the forces generated on the ends of the polymer due to flow. The chains then begin to stretch. This leads to anisotropy and stress differences between the elongation and compression planes, and is observed as an increase in the elongational viscosity in Equation $\eta_E = \dfrac{\sigma_{xx} - \sigma_{yy}}{\dot{\varepsilon}} = \dfrac{\sigma_{xx} - \sigma_{zz}}{\dot{\varepsilon}}$. Since the polymers are not infinitely extensible, the stress difference between the planes saturates to the value given by that in the fully stretched state (which is like a slender rod).

The behaviour of elongational viscosity in concentrated solutions and melts is different. The zero shear rate viscosity is itself much higher than the solvent viscosity. A semi-quantitative model for the behaviour in linear polymer melts can be interpreted in terms of the tube model or also called as the "standard molecular theory". At strain rates small compared to the inverse reptation time (or disentanglement time) $\tau \dfrac{-1}{d}$, thermal fluctuations are stronger than the hydrodynamic drag forces induced by the flow, and the chain remain in close to equilibrium microstructure. The viscosity is nearly same as the zero-shear rate viscosity, as shown schematically in figure. At higher strain rates $\dot{\varepsilon} \ll \tau \dfrac{-1}{d}$, the tubes (the pervaded volume of the polymers), are disentangled and begin to orient along the extensional direction, which decreases the viscosity. This continues till the strain rate becomes comparable to the inverse Rouse relaxation time $\tau \frac{-1}{R}$, when chain stretching begins (similar to the case in dilute polymer solutions). This causes an increase in the viscosity. The terminal region of constant viscosity is occurs due to the chains attaining their maximum stretch. Most of these theoretical predictions have not been verified by experiments, mainly due to the difficulty in consistently measuring the elongational viscosity at very large strain rates and strains.

Viscoelasticity

Solids, fluids and gases. We are used to them in our day to day life. We breath the air, drink the water and eat the apple. We know that below 0 degrees Celsius water is frozen into ice. At room temperature water is a fluid and above 100 degrees Celsius it vaporizes. We live with the idea that matter is either a gas or a fluid or a solid.

Yet there are many materials that cannot be so easily classified. These materials can behave as a fluid but also as an elastic solid. We call these materials viscoelastic materials because, at the same time, they have both fluid (viscous) properties and elastic properties.

What is causing these viscoelastic properties? The answer is astonishing simple: any material that consists of long flexible fibres like particles is in nature viscoelastic. Because of their shape the particles can temporarily connect to each other which cause the elastic properties. On the other hand, due to their flexibility, they will easily slide along each other which cause the fluid properties.

Typical examples of viscoelastic materials are spaghetti, shag (tobacco), a pile of worms moving through each other and (of course) polymers. Polymers are always viscoelastic because they consist out of long molecules which can be entangled with their neighbors.

Young's Modulus of Elasticity

Young's Modulus

Imagine a piece of dough. Stretch it. It gets longer and thinner. Squash it. It gets shorter and fatter. Now imagine a piece of granite. Try the same mental experiment. The change in shape must surely occur, but to the unaided eye it's imperceptible. Some materials stretch and squash quite easily. Some do not.

The quantity that describes a material's response to stresses applied normal to opposite faces is called Young's modulus in honor of the English scientist Thomas Young. Young was the first person to define work as the force displacement product, the first to use the word energy in its modern sense, and the first to show that light is a wave. He was not the first to quantify the resistance of materials to tension and compression, but he became the most famous early proponent of the modulus that now bears his name. Young didn't name the modulus after himself. He called it the *elastic modulus*. The symbol for Young's modulus is usually E from the French word *élasticité* (elasticity) but some prefer Y in honor of the scientist.

Young's modulus is defined for all shapes and sizes by the same rule, but for convenience sake let's imagine a rod of length ℓ_0 and cross sectional area A being stretched by a force F to a new length $\ell_0 + \Delta\ell$.

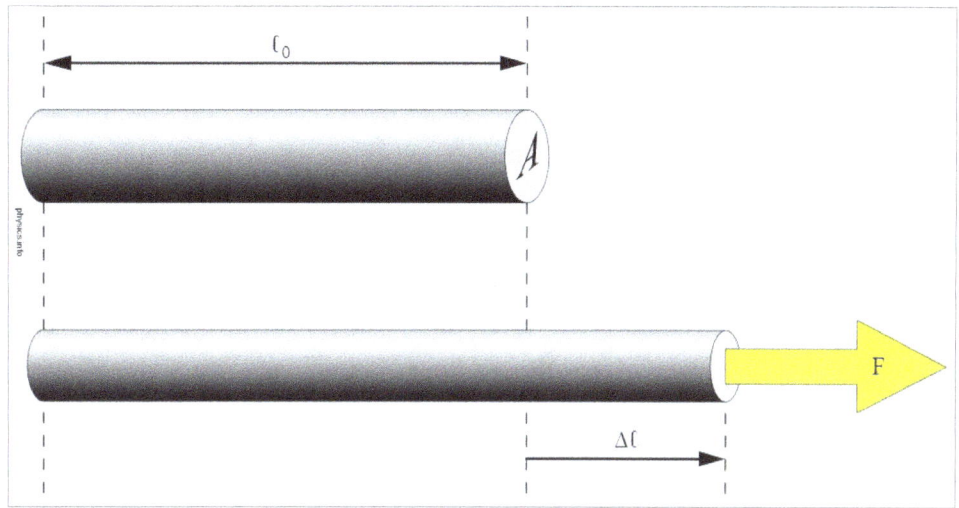

Young's modulus (extension)

Tensile stress is the outward normal force per area ($\sigma = F/A$) and *tensile strain* is the fractional increase in length of the rod ($\varepsilon = \Delta\ell/\ell_0$). The proportionality constant that relates these two quantities together is the ratio of tensile stress to tensile strain —*Young's modulus*.

$$\frac{F}{A} = E\frac{\Delta\ell}{\ell} \qquad\qquad \sigma = E\varepsilon$$

The same relation holds for forces in the opposite direction; that is, a strain that tries to shorten an object.

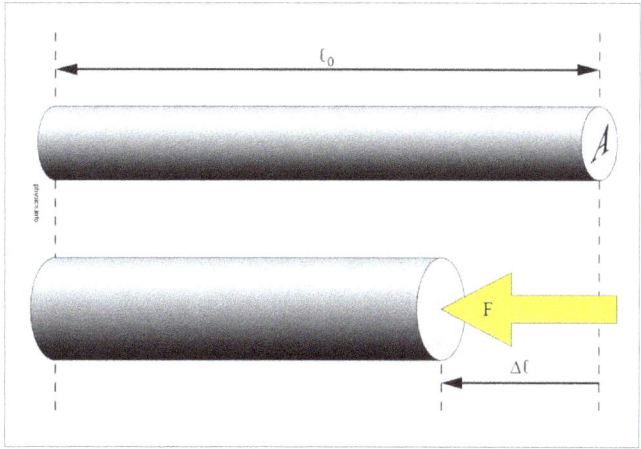

Young's Modulus (Compression)

Replace the adjective tensile with compressive. The normal force per area directed inward ($\sigma = F/A$) is called the *compressive stress* and the fractional decrease in length ($\varepsilon = \Delta\ell/\ell_0$) is called the *compressive strain*. This makes *Young's modulus* the ratio of compressive stress to compressive strain. The adjective may have changed, but the mathematical description did not.

$$\frac{F}{A} E \frac{L\ell}{\ell_0} \qquad \sigma = E\varepsilon$$

The SI units of Young's modulus is the *pascal* [Pa].

$$\left[\frac{N}{A} = Pa\,\frac{m}{m}\right]$$

but for most materials the *gigapascal* is more appropriate [GPa].

1 GPa = 109 Pa.

References

- Thermodynamics, science: britannica.com, Retrieved 5 June, 2019

- Thermodynamics-of-Mixing, Ideal-Systems, Thermodynamics, Physical-and-Theoretical-Chemistry: libretexts. org, Retrieved 8 August, 2019

- What-is-viscoelasticity: viscoelasticity.info, Retrieved 11 January, 2019

- Elasticity: physics.info, Retrieved 14 July, 2019

Permissions

Index